North American Aviation, Inc.

North American X-15 Rocket Plane Pilot's
Flight Operating Instructions

©2006-2010 Periscope Film LLC
All Rights Reserved
ISBN # 978-1-935327-86-8   1-935327-86-0

International Airport,   Los Angeles 45, California
LOS ANGELES DIVISION

T.O. 1X-15-1
(FORMERLY FHB-23-1)

UTILITY FLIGHT MANUAL

USAF SERIES

X-15

AIRCRAFT

CONTRACT
AF33(600)-31693
AF33(600)-40765

Commanders are responsible for bringing this
publication to the attention of all personnel
cleared for operation of subject aircraft.

PUBLISHED UNDER AUTHORITY OF THE
SECRETARY OF THE AIR FORCE

This publication replaces FHB-23-1 dated 18
March 1960. See Safety of Flight Supplement
Index, T.O. 0-1-1A, for current status of Safety
of Flight Supplements.

©2006-2010 Periscope Film LLC
All Rights Reserved
ISBN #978-1-935327-86-8 1-935327-86-0

29 DECEMBER 1961
CHANGED 26 APRIL 1963

T.O. 1X-15-1
(FORMERLY FHB-23-1)

UTILITY FLIGHT MANUAL

USAF SERIES X-15 AIRCRAFT

CONTRACT
AF33(600)-31693
AF33(600)-40766

DECLASSIFIED

Commanders are responsible for bringing this publication to the attention of all personnel cleared for operation of subject aircraft.

PUBLISHED UNDER AUTHORITY OF THE
SECRETARY OF THE AIR FORCE

See Safety of Flight Supplement Index, T.O. 0-1-1A, for current status of Safety of Flight Supplements.

AIR FORCE Kayy Litho Los Angeles, Cal 16 May 63-150 (North American Aviation)

CHANGE NOTICE

LATEST CHANGED PAGES SUPERSEDE
THE SAME PAGES OF PREVIOUS DATE
Insert changed pages into basic
publication. Destroy superseded pages.

©2006-2010 Periscope Film LLC
All Rights Reserved
ISBN #978-1-935327-86-8   1-935327-86-0

29 DECEMBER 1961
CHANGED 26 APRIL 1963

T.O. 1X-15-1

Reproduction for nonmilitary use of the information or illustrations contained in this publication is not permitted without specific approval of the issuing service. The policy for use of Classified Publications is established for the Air Force in AFR 205-1.

INSERT LATEST CHANGED PAGES. DESTROY SUPERSEDED PAGES.

LIST OF EFFECTIVE PAGES

NOTE: The portion of the text affected by the changes is indicated by a vertical line in the outer margins of the page.

TOTAL NUMBER OF PAGES IN THIS PUBLICATION IS 128, CONSISTING OF THE FOLLOWING:

Page No.	Issue
*Title	26 Apr 63
*A	26 Apr 63
i	Original
ii Blank	Original
iii thru iv	Original
*1-1 thru 1-6	26 Apr 63
1-7	Original
*1-8 thru 1-10	26 Apr 63
1-11 thru 1-13	Original
*1-14 thru 1-15	26 Apr 63
1-16	Original
1-17 thru 1-19	29 Jun 62
*1-20	26 Apr 63
1-21	Original
1-22 thru 1-22A	29 Jun 62
1-22B Blank	29 Jun 62
1-23	Original
*1-24 thru 1-25	26 Apr 63
1-26 thru 1-28	Original
1-29 thru 1-30	29 Jun 62
*1-31 thru 1-32	26 Apr 63
1-33	Original
*1-34 thru 1-34A	26 Apr 63
1-34B	29 Jun 62
*1-35 thru 1-36	26 Apr 63
1-37	Original
*1-38 thru 1-40	26 Apr 63
1-41 thru 1-42	Original
1-43 thru 1-44	29 Jun 62
1-45 thru 1-46	Original
1-47	29 Jun 62
1-48 Blank	Original
2-1 thru 2-2	Original
*2-3 thru 2-4	26 Apr 63
2-5	29 Jun 62
2-6 thru 2-7	Original
*2-8	26 Apr 63
2-9 thru 2-10	Original
*2-11 thru 2-15	26 Apr 63
2-16	Original
*2-17	26 Apr 63
2-18	Original
*2-19	26 Apr 63
2-20 Blank	Original
3-1 thru 3-2	Original
*3-3	26 Apr 63
3-4	Original
*3-5	26 Apr 63
3-6 thru 3-10	Original
*3-11	26 Apr 63

Page No.	Issue
3-12 thru 3-14	Original
4-1	29 Jun 62
4-2 thru 4-4	Original
4-5 thru 4-6A	29 Jun 62
4-6B Blank	29 Jun 62
*4-7 thru 4-9	26 Apr 63
4-10	29 Jun 62
*4-11	26 Apr 63
4-12 Blank	Original
5-1	29 Jun 62
*5-2 thru 5-3	26 Apr 63
5-4 thru 5-15	Original
5-16 Blank	Original
6-1 thru 6-5	Original
6-6 Blank	Original

*The asterisk indicates pages changed, added, or deleted by the current change.

ADDITIONAL COPIES OF THIS PUBLICATION MAY BE OBTAINED AS FOLLOWS:

USAF ACTIVITIES.—In accordance with T.O. 00-5-2.

A-2
USAF

A

Changed 26 April 1963

TABLE OF CONTENTS

Section	I	DESCRIPTION	1-1
Section	II	NORMAL PROCEDURES	2-1
Section	III	EMERGENCY PROCEDURES	3-1
Section	IV	AUXILIARY EQUIPMENT	4-1
Section	V	OPERATING LIMITATIONS	5-1
Section	VI	FLIGHT CHARACTERISTICS	6-1

THIS PAGE INTENTIONALLY LEFT BLANK.

T.O. 1X-15-1

This utility Flight Manual is the result of extensive research and analysis of engineering data. It contains the necessary information for safe and efficient operation of the X-15 Airplane. Information involving safety of flight will be disseminated by means of the regular Safety of Flight Supplement program. You can determine the status of Safety of Flight Supplements by referring to the Safety of Flight Supplement Index, T.O. 0-1-1A. The title page of the Flight Manual and title block of each Safety of Flight Supplement should also be checked to determine the effect that these publications may have on existing Safety of Flight Supplements. The manual is divided into six separate sections, each containing its own table of contents. The research program for which this airplane was designed requires that each individual mission be precisely preplanned. Consequently, standard performance data is not included in this manual. The Flight Manual does not discuss in detail certain complex units installed in the airplane, nor does it necessarily contain information on the use or operation of test equipment.

X-15 (Three-View)

Description

SECTION I

TABLE OF CONTENTS	PAGE		PAGE
Airplane	1-1	Launch System	1-36
Engine	1-3	Landing Gear System	1-37
Propellant Supply System	1-15	Instruments	1-37
Engine and Propellant Control Helium System	1-16	Inertial All-attitude Flight Data System (Gyro-stabilized Platform)	1-38
Propellant Pressurization Helium System	1-16	Instrumentation System	1-40
Auxiliary Power Units	1-17	Indicator, Caution, and Warning Light System	1-41
Electrical Power Supply Systems	1-22		
Hydraulic Power Supply Systems	1-26	Canopy	1-41
Flight Control Systems	1-27	Ejection Seat	1-43
Wing Flap System	1-36		
Speed Brake System	1-36	Auxiliary Equipment	1-46

AIRPLANE.

The X-15 is a single-place research airplane, specifically designed to obtain data on flight at extremely high altitudes and speeds and on the physiological and psychological effects of such flight conditions on the pilot. Built by North American Aviation, Inc, the airplane has an inertial all-attitude (gyro-stabilized platform) flight data system and is powered by one XLR99 liquid-propellant rocket engine. The 25-1/2 degree swept-back wing has hydraulically operated flaps on the inboard trailing edge of each wing panel. All aerodynamic control surfaces are actuated by irreversible hydraulic systems. The horizontal stabilizer has a 15-degree cathedral. The two sections move simultaneously for pitch control, differentially for roll control, and in compound for pitch-roll control. The upper and lower vertical stabilizers are in two sections, a movable outer span for yaw control and a fixed section adjacent to the fuselage. The lower movable section (ventral) is jettisonable for landing. Each fixed section incorporates a split-flap speed brake. For changes in airplane attitude relative to flight trajectory at altitudes where aerodynamic controls are relatively ineffective, the airplane incorporates a ballistic control system, wherein the metered release of gas through small rockets in the nose and wing causes the airplane to move about each axis as required. Two auxiliary power units drive the airplane hydraulic pumps and ac electrical generators. Fuel for the rocket engine is carried internally. The airplane is not designed for normal ground take-off, but is air-launched by a B-52 Airplane. The landing gear consists of a dual-wheel nose gear and two main landing skids. The gear is lowered in flight by gravity and air loads.

AIRPLANE DIMENSIONS.

The over-all dimensions of the airplane (in-flight configuration with gear up and ventral retained) are as follows:

Length .... 49 feet 2 inches
Span .... 22 feet 4 inches
Height .... 13 feet 1 inch

NOTE

In the landing configuration (landing gross weight and gear down, with specified nose tire and strut inflation and with ventral jettisoned), height is 11 feet 6 inches.

AIRPLANE GROSS WEIGHT.

The approximate launch gross weight of the airplane (including full internal load and pilot) is 32,900 pounds. However, this can vary a few hundred pounds, depending on the type of instrumentation carried.

Section I
T.O. 1X-15-1

GENERAL ARRANGEMENT

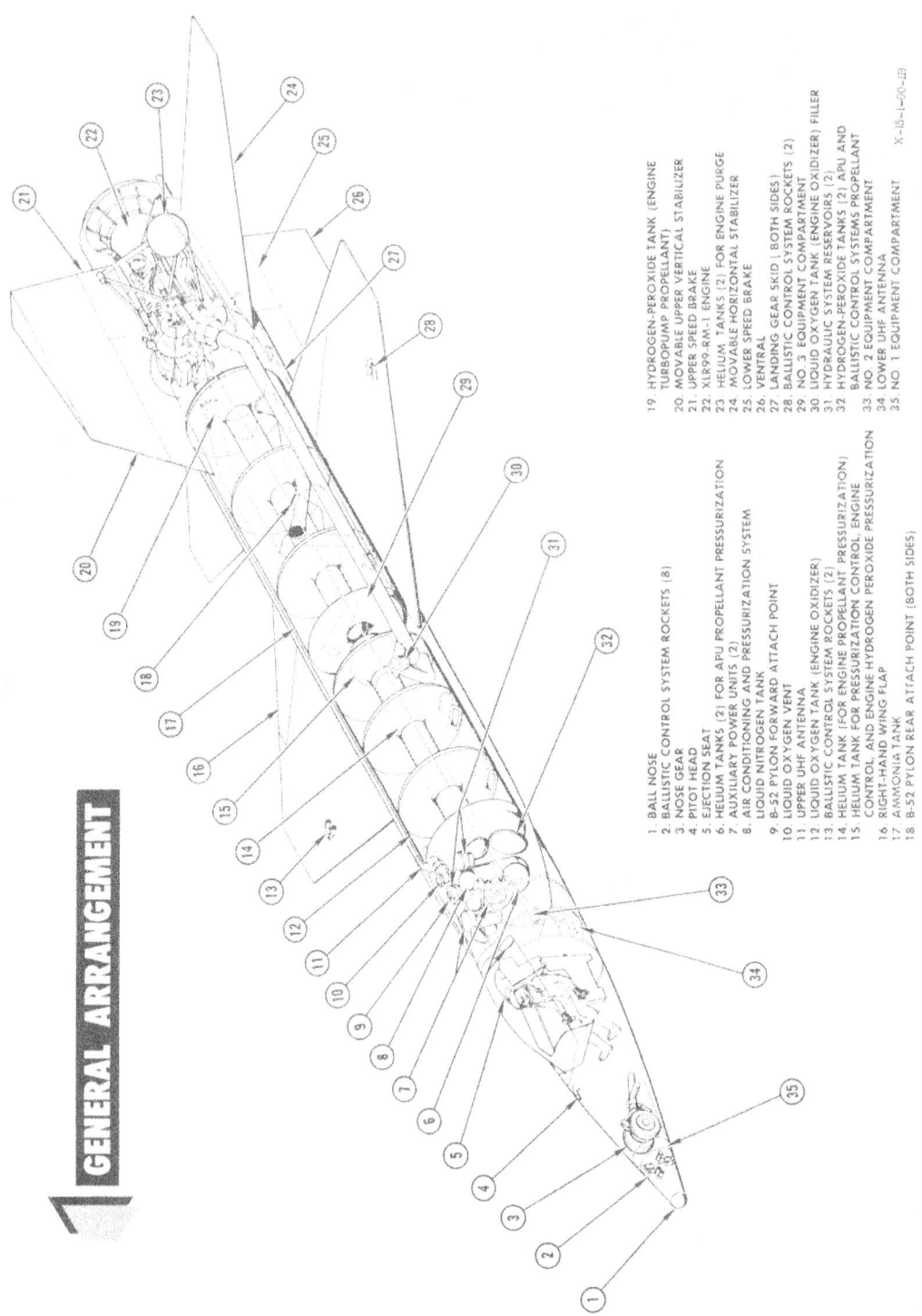

1. BALL NOSE
2. BALLISTIC CONTROL SYSTEM ROCKETS (8)
3. NOSE GEAR
4. PITOT HEAD
5. EJECTION SEAT
6. HELIUM TANKS (2) FOR APU PROPELLANT PRESSURIZATION
7. AUXILIARY POWER UNITS (2)
8. AIR CONDITIONING AND PRESSURIZATION SYSTEM LIQUID NITROGEN TANK
9. B-52 PYLON FORWARD ATTACH POINT
10. LIQUID OXYGEN VENT
11. UPPER UHF ANTENNA
12. LIQUID OXYGEN TANK (ENGINE OXIDIZER)
13. BALLISTIC CONTROL SYSTEM ROCKETS (2)
14. HELIUM TANK (FOR ENGINE PROPELLANT PRESSURIZATION)
15. HELIUM TANK FOR PRESSURIZATION CONTROL ENGINE CONTROL, AND ENGINE HYDROGEN PEROXIDE PRESSURIZATION
16. RIGHT-HAND WING FLAP
17. AMMONIA TANK
18. B-52 PYLON REAR ATTACH POINT (BOTH SIDES)
19. HYDROGEN-PEROXIDE TANK (ENGINE TURBOPUMP PROPELLANT)
20. MOVABLE UPPER VERTICAL STABILIZER
21. UPPER SPEED BRAKE
22. XLR99-RM-1 ENGINE
23. HELIUM TANKS (2) FOR ENGINE PURGE
24. MOVABLE HORIZONTAL STABILIZER
25. LOWER SPEED BRAKE
26. VENTRAL
27. LANDING GEAR SKID (BOTH SIDES)
28. BALLISTIC CONTROL SYSTEM ROCKETS (2)
29. NO. 3 EQUIPMENT COMPARTMENT
30. LIQUID OXYGEN TANK (ENGINE OXIDIZER) FILLER
31. HYDRAULIC SYSTEM RESERVOIRS (2)
32. HYDROGEN-PEROXIDE TANKS (2) APU AND BALLISTIC CONTROL SYSTEMS PROPELLANT
33. NO. 2 EQUIPMENT COMPARTMENT
34. LOWER UHF ANTENNA
35. NO. 1 EQUIPMENT COMPARTMENT

Figure 1-1

AIRPLANE SERIAL NUMBERS.

The Air Force serial numbers for X-15 Airplanes covered by this manual are AF56-6670, -6671, and -6672.

ENGINE.

Thrust is provided by one YLR99 turborocket engine. It has a single thrust chamber, a two-stage, continuous-igniter starting system, a turbopump, and a gas generator. Propellants are liquid oxygen and anhydrous ammonia, supplied from the airplane propellant system. (See figure 1-6.) The engine is of variable-thrust design, capable of operating over a range of 30 to 100 percent of full rated thrust. For this airplane, however, the variable thrust range will normally be 50 to 100 percent. For specific missions, a range of 40 to 100 percent is available. The throttle quadrant markings serve to identify the thrust range capability being used. The gas generator decomposes a monopropellant fuel, 90 percent hydrogen peroxide (H_2O_2), to provide a high-pressure gas mixture for driving the turbopump, which in turn drives the two centrifugal pumps that supply the propellants to the engine. Upon discharge from the pumps, the propellants are delivered to the two igniters and the thrust chamber where they are burned. At the first-stage igniter, the oxygen (in gaseous form) and ammonia are mixed and then ignited by three spark plugs. Liquid oxygen and ammonia, meanwhile, also are routed to the second-stage igniter. When the pressure created by the hot gases in the first-stage igniter actuates a pressure switch, propellants are allowed to enter the second-stage igniter. Here, the propellants are mixed and ignited by the incoming first-stage gases at greatly increased pressure. When a pressure switch in the second-stage igniter is actuated, propellants are allowed to enter the thrust chamber itself. They are again mixed and ignited by the gases coming from the second-stage igniter and build to the tremendous pressures needed for required thrust. The thrust chamber is an assembly of small welded, wire-wound tubes preformed as segments of the chamber. Before injection into the thrust chamber, the ammonia passes through these tubes to cool the chamber. Exhaust gases are discharged through a venturi-shaped sonic nozzle.

ENGINE COMPARTMENT.

The engine compartment, in the extreme aft end of the fuselage, houses the tubular steel engine mount which supports the engine and turbopump. The engine compartment is completely isolated from the airframe by a mono-fire-wall. A large access door is provided in the forward end of the engine compartment for access to the engine compartment from the hydrogen peroxide storage tank area. The engine compartment also houses the instrumentation pick-offs, fire detection system sensors, and helium release line. A fire seal closes out the compartment and protects against the entry of exhaust gases and expelled propellants into the engine compartment. For engine compartment purging, refer to "Engine Compartment Purging System" in this section.

Engine Compartment Fire Detection System.

A detection circuit is provided to detect and indicate a fire condition in the engine compartment. This circuit is of the continuous-element type, which detects excessive temperatures anywhere along its length. The system is powered by the battery bus and continuously monitors the resistance of the circuit. The resistance of the material used in the circuit varies inversely with temperature and total length of the sensing circuit. Whenever temperatures in the engine compartment reach 1100°F (594°C) or higher, the resistance of the sensing element falls below a preset value because of the excessive temperature, and the warning system is energized. A placard-type warning light, a system selector switch, and a test switch are in the cockpit. For emergency procedures in case of a fire-warning indication, refer to "Fire or Explosion" in Section III.

Fire-warning Light. An abnormal rise in engine compartment temperature is shown by a placard-type warning light (70, figure 1-2), on the instrument panel. The light is powered by the primary dc bus and has a red plastic cap which shows the word "FIRE" when the light is on. The light may be tested by a push-to-test switch on the instrument panel right wing.

Fire-warning Light Test Button. A fire-warning light and detection circuit test button (30, figure 1-2) is on the instrument panel right wing. The button is powered by the primary dc bus. When the button is pressed, the fire-warning light should come on, verifying the continuity of the detection circuit.

Engine Compartment Purging System.

The engine compartment can be purged by releasing an inert gas (helium) under pressure into the area to extinguish a fire or relieve an overheat condition. Three cubic feet of helium is stored in two spherical containers under 3600 psi pressure. The containers are on either side of the engine compartment in the left and right wing root fairing tunnels. Either automatic or manual release of helium can be selected by the pilot. Because of the location of the two containers adjacent to the engine compartment, any high-temperature condition in the compartment will affect these containers and create a potential explosion hazard. As the helium is released into the engine compartment, it inhibits any fire condition and at the same time eliminates the explosion hazard.

Helium Release Selector Switch. This three-position switch (69, figure 1-2), labeled "HE REL SW," is on the left side of the instrument panel. It permits the pilot to select the type of engine compartment purging (either automatic or manual) in case of a fire indication. The switch is powered by the battery bus. The AUTO position sets up an entirely automatic sequence if a fire occurs in the engine compartment, as indicated by illumination of the fire-warning light. The engine is shut down, and the helium from the two containers outboard of the engine is jettisoned into the engine compartment to inhibit the fire or overheat condition and to prevent overpressurization of the containers by extreme temperature increase. The OFF position sets up the fire detection system for illuminating the fire-warning light only, in case a fire occurs. It will then be necessary either to move the switch to ON to release the helium into the engine compartment without affecting

Section I
T.O. 1X-15-1

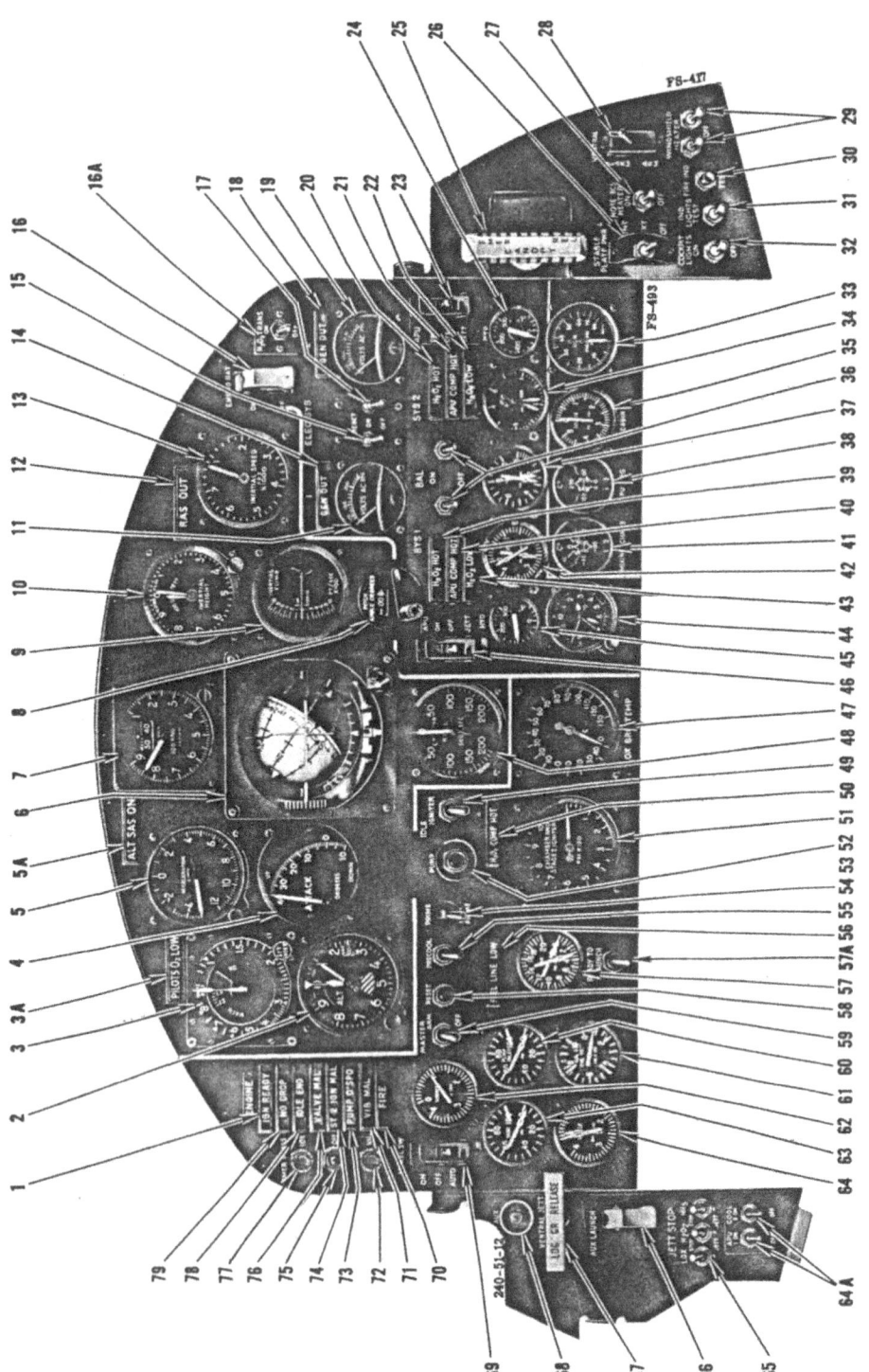

Figure 1-2

1-4
CHANGED 26 APRIL 1963

1. IGNITION-READY CAUTION LIGHT
2. ALTIMETER
3. AIRSPEED/MACH INDICATOR
3A. PILOT'S OXYGEN-LOW CAUTION LIGHT
4. ANGLE-OF-ATTACK INDICATOR
5. ACCELEROMETER
5A. ALTERNATE SAS ON LIGHT*
6. ATTITUDE INDICATOR
7. AIRSPEED INDICATOR
8. PITCH ANGLE SET CONTROL
9. VERTICAL VELOCITY INDICATOR
10. INERTIAL HEIGHT (ALTIMETER) INDICATOR
11. NO. 1 GENERATOR VOLTMETER
12. RAS-OUT INDICATOR LIGHT
13. INERTIAL SPEED (VELOCITY) INDICATOR
14. NO. 1 GENERATOR-OUT LIGHT
15. NO. 1 GENERATOR SWITCH
16. EMERGENCY BATTERY SWITCH
16A. HYDROGEN PEROXIDE TRANSFER SWITCH
17. NO. 2 GENERATOR SWITCH
18. NO. 2 GENERATOR-OUT LIGHT
19. NO. 2 GENERATOR VOLTMETER
20. NO. 2 APU HYDROGEN PEROXIDE OVERHEAT WARNING LIGHT
21. NO. 2 APU COMPARTMENT OVERHEAT CAUTION LIGHT
22. NO. 2 APU HYDROGEN PEROXIDE-LOW CAUTION LIGHT
23. NO. 2 APU SWITCH
24. NO. 2 HYDRAULIC TEMPERATURE GAGE
25. CANOPY INTERNAL EMERGENCY JETTISON HANDLE
26. STABLE PLATFORM SWITCH
27. NOSE BALLISTIC ROCKET HEATER SWITCH
28. VENTRAL ARMING SWITCH
29. WINDSHIELD HEATER SWITCHES (2)
30. FIRE-WARNING LIGHT TEST BUTTON
31. INDICATOR, CAUTION, AND WARNING LIGHT SWITCH
32. COCKPIT LIGHTING SWITCH
33. CABIN PRESSURE ALTIMETER
34. HYDRAULIC PRESSURE GAGE
35. CABIN HELIUM SOURCE PRESSURE GAGE
36. NO. 1 AND NO. 2 BALLISTIC CONTROL SWITCHES
37. HYDROGEN PEROXIDE TANK PRESSURE GAGE
38. APU BEARING TEMPERATURE GAGE
39. NO. 1 APU HYDROGEN PEROXIDE OVERHEAT WARNING LIGHT
40. NO. 1 APU COMPARTMENT OVERHEAT CAUTION LIGHT
41. MIXING CHAMBER TEMPERATURE GAGE
42. APU SOURCE PRESSURE GAGE
43. NO. 1 APU HYDROGEN PEROXIDE-LOW CAUTION LIGHT
44. CLOCK
45. NO. 1 HYDRAULIC TEMPERATURE GAGE
46. NO. 1 APU SWITCH
47. LIQUID OXYGEN BEARING TEMPERATURE GAGE
48. RATE-OF-ROLL INDICATOR
49. IGNITER IDLE SWITCH
50. H_2O_2 COMPARTMENT-HOT CAUTION LIGHT
51. CHAMBER AND STAGE 2 IGNITER PRESSURE GAGE
52. TURBOPUMP IDLE BUTTON
53. (DELETED)
54. ENGINE PRIME SWITCH
55. ENGINE PRECOOL SWITCH
56. FUEL LINE-LOW CAUTION LIGHT
57. PROPELLANT MANIFOLD PRESSURE GAGE
57A. READY-TO-LAUNCH SWITCH
58. ENGINE RESET BUTTON
59. ENGINE MASTER SWITCH
60. PROPELLANT PUMP INLET PRESSURE GAGE
61. H_2O_2 TANK AND ENGINE CONTROL LINE PRESSURE GAGE
62. PROPELLANT SOURCE PRESSURE GAGE
63. PROPELLANT TANK PRESSURE GAGE
64. H_2O_2 SOURCE AND PURGE PRESSURE GAGE
64A. APU WATER COOLING SWITCHES
65. JETTISON STOP SWITCHES
66. AUXILIARY LAUNCH SWITCH
67. LANDING GEAR HANDLE
68. VENTRAL JETTISON BUTTON
69. HELIUM RELEASE SELECTOR SWITCH
70. FIRE-WARNING LIGHT
71. ENGINE VIBRATION MALFUNCTION CAUTION LIGHT
72. AMMONIA TANK PRESSURE-LOW CAUTION LIGHT
73. TURBOPUMP OVERSPEED CAUTION LIGHT
74. STAGE 2 IGNITION MALFUNCTION CAUTION LIGHT
75. PROPELLANT EMERGENCY PRESSURIZATION SWITCH
76. VALVE MALFUNCTION CAUTION LIGHT
77. LIQUID OXYGEN TANK PRESSURE-LOW CAUTION LIGHT
78. IDLE END LIGHT
79. NO-DROP CAUTION LIGHT

* Some airplanes (Refer to applicable text.)

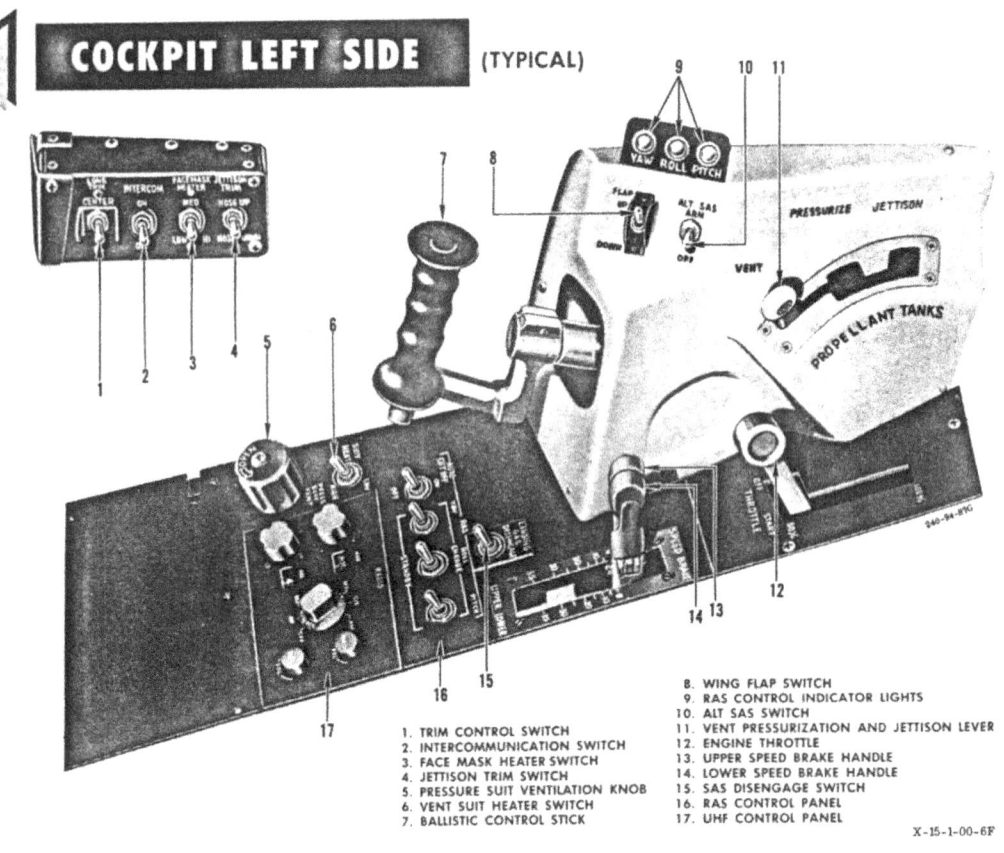

Figure 1-3

engine operation, or to move the switch to AUTO to release the helium and simultaneously shut down the engine. Moving the switch to ON will release the helium into the engine compartment whenever the battery bus is energized. Once energized, the helium release valve is locked in the jettison position and must be electrically unlocked by ground personnel.

ENGINE TURBOPUMP.

A turbopump, mounted on the front of the engine, delivers the propellants in the desired quantities and at the proper pressures to the engine. The turbopump contains a gas generator, a two-stage, axial-flow turbine, and two centrifugal pumps on a common shaft. Each pump supplies one of the propellants to the engine. The monopropellant for driving the turbine is 90 percent hydrogen peroxide (H_2O_2). An electrohydraulic power servo system governs turbine speed to the selected power requirement.

Turbopump Propellant (H_2O_2) System.

The hydrogen peroxide monopropellant (H_2O_2) used to drive the turbopump is contained in a 10-cubic-foot spherical supply tank (19, figure 1-1) with a capacity of 854 pounds (77.5 US gallons). A swivel-type pickup feed line allows positive feeding of the monopropellant regardless of airplane attitude. The system includes a combination vent, pressure relief, and tank pressurization valve; a jettison valve; a hydrogen peroxide throttle control metering valve; a safety valve; a shut-off valve; and a gas generator. This system is controlled by switches and a control lever in the cockpit and is put into operation whenever the engine starting sequence is begun. For a description of these controls, refer to "Engine Controls" in this section. When the engine is not operating, the tank is vented to atmosphere if the vent, pressurization, and jettison control lever is at VENT and control gas is available. The tank is pressurized with helium control gas, to feed the H_2O_2 to the gas generator, which provides steam power for turbopump operation. Tank pressure can be read from a gage in the cockpit. Refer to "H_2O_2 Tank and Engine Control Line Pressure Gage" in this section. The system also includes a jettison feature that permits the H_2O_2 to be forcibly expelled overboard.

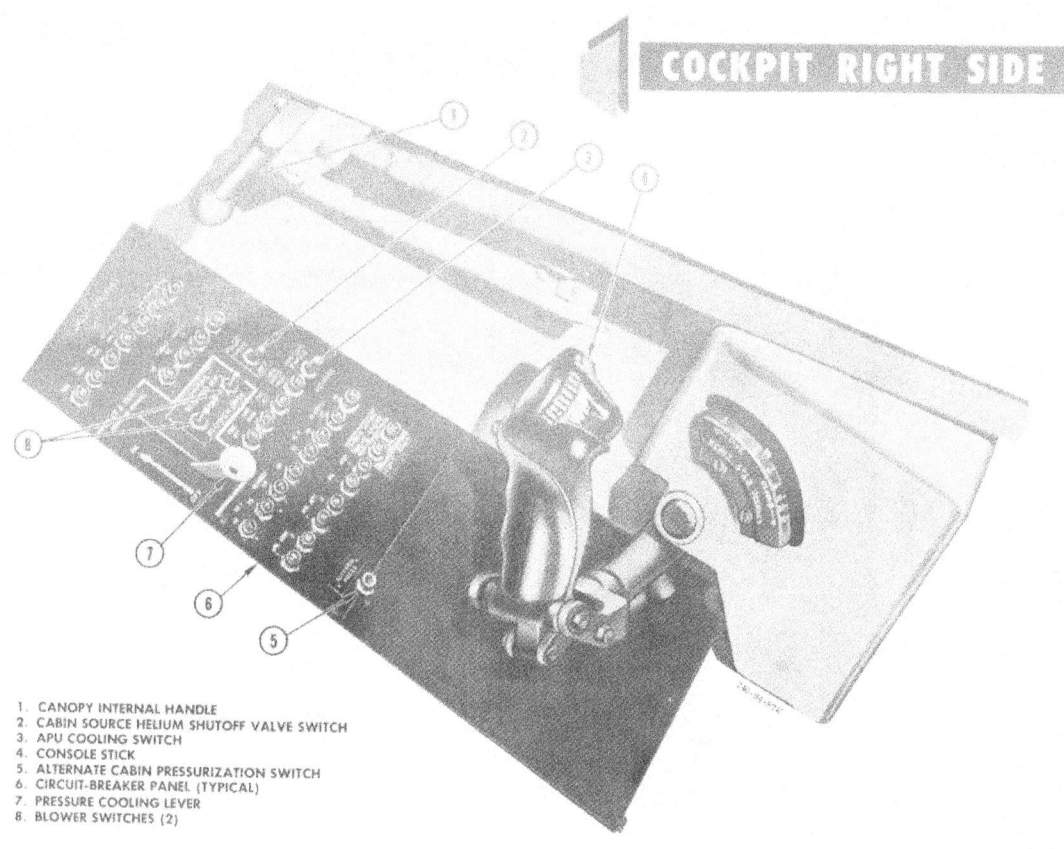

1. CANOPY INTERNAL HANDLE
2. CABIN SOURCE HELIUM SHUTOFF VALVE SWITCH
3. APU COOLING SWITCH
4. CONSOLE STICK
5. ALTERNATE CABIN PRESSURIZATION SWITCH
6. CIRCUIT-BREAKER PANEL (TYPICAL)
7. PRESSURE COOLING LEVER
8. BLOWER SWITCHES (2)

Figure 1-4.

H_2O_2 Compartment-hot Light. An amber H_2O_2 compartment-hot caution light (50, figure 1-2), on the instrument panel, comes on when temperature in the upper area of the turbopump propellant compartment reaches 538°C (1000°F) or when temperature in the lower area of the compartment reaches 427°C (800°F). When illuminated, the light reads "H_2O_2 COMP HOT." The light is powered by the primary dc bus and may be tested through the indicator, caution, and warning light test circuit.

Turbopump Speed Control.

An electrohydraulic servo system is used as an actuation and reference system between the turbine speed and H_2O_2 flow. Its main components are a power package, a governor, a throttle synchro, a servo amplifier, a governor actuator, and an H_2O_2 throttle control metering valve. Pressurized oil from an electrically driven hydraulic pump is supplied to the governor and metering valve. When the engine throttle (throttle synchro) is moved, the governor speed adjustment lever is set to the desired position by the governor actuator. The speed of the turbopump is sensed by the governor, and the hydraulic pressure balance between the governor and metering valve is adjusted to control peroxide flow into the gas generator. Decrease or increase of the turbopump speed from that required for the selected thrust causes a hydraulic imbalance between governor and metering valve. As the governor reacts to restore the hydraulic balance, hydraulic pressure to the metering valve is increased or decreased, as necessary, to alter the rate of H_2O_2 flow to the gas generator and thus restore the turbopump to the desired speed.

ENGINE PROPELLANT AND CONTROL SYSTEM.

The two propellants, anhydrous ammonia and liquid oxygen, are routed from their respective fuel tanks to the main feed valves, which are operated by helium pressure. From the main feed valves, the fuel is routed to the turbopump. The fuels, pressurized by a low-pressure inert gas (helium), flow from the respective tanks to the turbopump. Prime orifices allow propellants to circulate and cool the engine and prime the propellant pumps. The turbopump begins operation when the hydrogen peroxide supply upstream safety valve and downstream shutoff valves are opened. The

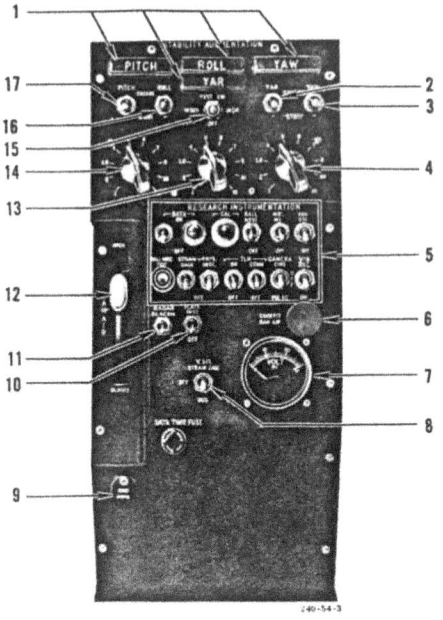

1. SAS CAUTION LIGHTS
2. YAR FUNCTION SWITCH
3. YAW FUNCTION SWITCH
4. YAW GAIN SELECTOR KNOB
5. INSTRUMENTATION CONTROL PANEL
6. COCKPIT RAM-AIR KNOB
7. DC VOLTMETER
8. DC VOLTMETER SWITCH
9. GROUND INTERPHONE RECEPTACLE
10. STABLE PLATFORM INSTRUMENT SWITCH
11. RADAR BEACON SWITCH
12. RAM-AIR LEVER
13. ROLL AND YAR GAIN SELECTOR KNOB
14. PITCH GAIN SELECTOR KNOB
15. SAS TEST SWITCH
16. ROLL FUNCTION SWITCH
17. PITCH FUNCTION SWITCH

Figure 1-5

H_2O_2 then flows to a gas generator, where it is converted to a high-pressure gas mixture of superheated steam and oxygen to drive the turbine wheel, which in turn drives the propellant pumps. The propellants are then supplied to the first-stage and second-stage igniters and to the main thrust chamber. After priming is completed, the turbopump is operating, and the first-stage igniter propellant valve is opened, the liquid oxygen to the first-stage igniter is routed inside the turbine exhaust duct, whose hot gases heat the liquid oxygen and change it to a gaseous state. The gaseous oxygen and ammonia then enter the first-stage igniter. Three spark plugs in the first-stage igniter fire the fuel and oxidizer mixture. When the pressure switch in the first-stage igniter is actuated, the second-stage igniter start valves open, allowing liquid oxygen and ammonia to flow into the second-stage igniter. First-stage igniter flames ignite the second-stage fuel mixture. Combustion pressure in the second-stage igniter then actuates a switch which signals the main propellant valve to open. When the main propellant valve opens, fuel and liquid oxygen are injected into the main thrust chamber, where they are ignited by second-stage flames. Before entering the main thrust chamber, the ammonia is routed through the chamber tubes in order to cool the main thrust chamber. Opening of the main propellant valve stops the flow of propellants to the prime valves. Once engine operation has been initiated, thrust output is varied between 50% and 100% according to the throttle position selected by the pilot. The engine propellant control system is shown schematically in figure 1-7.

ENGINE CONTROLS.

Throttle.

The throttle (12, figure 1-3) controls thrust output of the engine. The throttle quadrant has four marked positions: OFF, START, 50%, and 100%. When the throttle linkage is modified to allow a lower minimum thrust limit, the markings are OFF, START, 40%, and 100%. The throttle controls an electromechanical servo system, which includes a synchro transmitter attached to the throttle, a servo amplifier, and an actuator position transmitting synchro linked to the turbopump governor. Turbopump and first- and second-stage igniter operation is accomplished with the throttle at OFF (full aft and outboard). During the 30-second idle operation period with the throttle OFF, the turbopump is automatically maintained at idle speed. Within 30 seconds after igniter idle operation is begun, the throttle must be moved toward START and set at 50% (40%) to open the main propellant valves to the main thrust chamber or the start sequence must be terminated. After main thrust chamber operation is begun, movement of the throttle will vary engine thrust accordingly.

Vent, Pressurization, and Jettison Lever.

This lever (11, figure 1-3) controls the pressurization system selector valve. The valve is a manually controlled pneumatic selector valve. The lever has three positions: VENT, PRESSURIZE, and JETTISON. With the lever at VENT, helium pressure (from the helium pressure control system) is applied to all tank control valves in the propellant system. The pressurization valves close and the vent valves open, venting the H_2O_2, liquid oxygen, and ammonia tanks. In order to obtain engine operation, the lever must be moved to PRESSURIZE, opening the propellant system pressurization valves and closing the vent valves. This allows helium to enter and pressurize the turbopump H_2O_2 supply tank and the liquid oxygen and ammonia tanks. When the lever is positioned to JETTISON, helium pressure is applied to open three jettison valves and pressurize the H_2O_2, liquid oxygen, and ammonia tanks. The three propellants will then begin to dump overboard.

NOTE

The propellants will not jettison if the jettison test switches are OFF.

Jettison Trim Switch.

This switch (4, figure 1-3) is on the left vertical side panel. It has three positions: NOSE UP, NOSE DOWN, and an unmarked, center off position. The switch is powered by the 28-volt primary dc bus. With the vent, pressurization, and jettison lever at JETTISON, the jettison stop switches in the JETT position, and this switch at the unmarked off position, simultaneous jettisoning of H_2O_2, liquid oxygen, and ammonia will occur. Moving this switch to NOSE UP (when nose-down trim is felt) stops the flow of the ammonia. Moving the switch to NOSE DOWN (when nose-up trim is felt) stops the flow of liquid oxygen. In either case, when the airplane returns to trim, the switch must be released and allowed to return to the unmarked off position.

Jettison Stop Switches.

Three jettison stop switches (65, figure 1-2), on the instrument panel left wing, have a STOP and a JETT position. These switches, powered by the primary dc bus, are normally left in the STOP position until the prelaunch cruise portion of the flight. To perform a test of the turbopump H_2O_2, liquid oxygen, and ammonia jettison system, the vent, pressurization, and jettison control lever should be placed at JETTISON. The systems then can be tested by placing the switches to JETT. The jettison line of each system should then emit a vaporous cloud. Flow will cease when the switches are returned to the STOP position or when the vent, pressurization, and jettison control lever is moved to PRESSURIZE. (See figure 1-15 for location of jettison, drain, and bleed outlets.)

Engine Master Switch.

The engine master switch (59, figure 1-2), on the instrument panel, is powered by the primary dc bus. With the switch at OFF, primary dc bus power for engine control and engine indicator lights is interrupted. With the switch at ARM, primary dc bus power is applied to the engine indicator lights and engine control switching units.

Engine Reset Button.

The engine reset button (58, figure 1-2), on the instrument panel, is powered by the primary dc bus through the engine master switch. For a normal engine start or if a malfunction causes automatic shutdown during any phase of operation, depressing this button positions the engine control circuits to the armed position. However, if the malfunction which caused shutdown persists, engine control circuits will not be armed.

Engine Precool Switch.

The engine precool switch (55, figure 1-2), on the instrument panel, is powered by the primary dc bus through the engine master switch. It has two maintained positions: PRECOOL and OFF (down). With an engine start sequence initiated, moving the switch to PRECOOL opens the liquid oxygen main feed valve and precools the system up to the main propellant valve. The precooling flow dumps overboard through the engine liquid oxygen prime valve.

NOTE

About 10 minutes is required to precool the engine liquid oxygen system. After precooling is completed, the engine can be maintained in a precooled condition for an extended period by the following schedule: engine precool switch at OFF for 20 minutes, then at PRECOOL for 7-1/2 minutes, repeating this cycle as often as necessary.

Engine Prime Switch.

The engine prime switch (54, figure 1-2), on the instrument panel, has three positions: an unmarked, maintained center position; a momentary PRIME position; and a maintained STOP PRIME position. With an engine start sequence initiated, moving the switch momentarily to PRIME opens the liquid oxygen and ammonia main feed valves and the turbopump H_2O_2 upstream safety valve and admits helium to the engine control and purge systems. Approximately 30 seconds is required for priming at high-flow rate, and when the engine precool switch is placed at OFF, prime continues at low-flow rate until an actual start stops the prime or until the engine prime switch is placed momentarily at STOP PRIME. Engine operation may be terminated during any phase by moving this switch to STOP PRIME.

Section I T.O. 1X-15-1

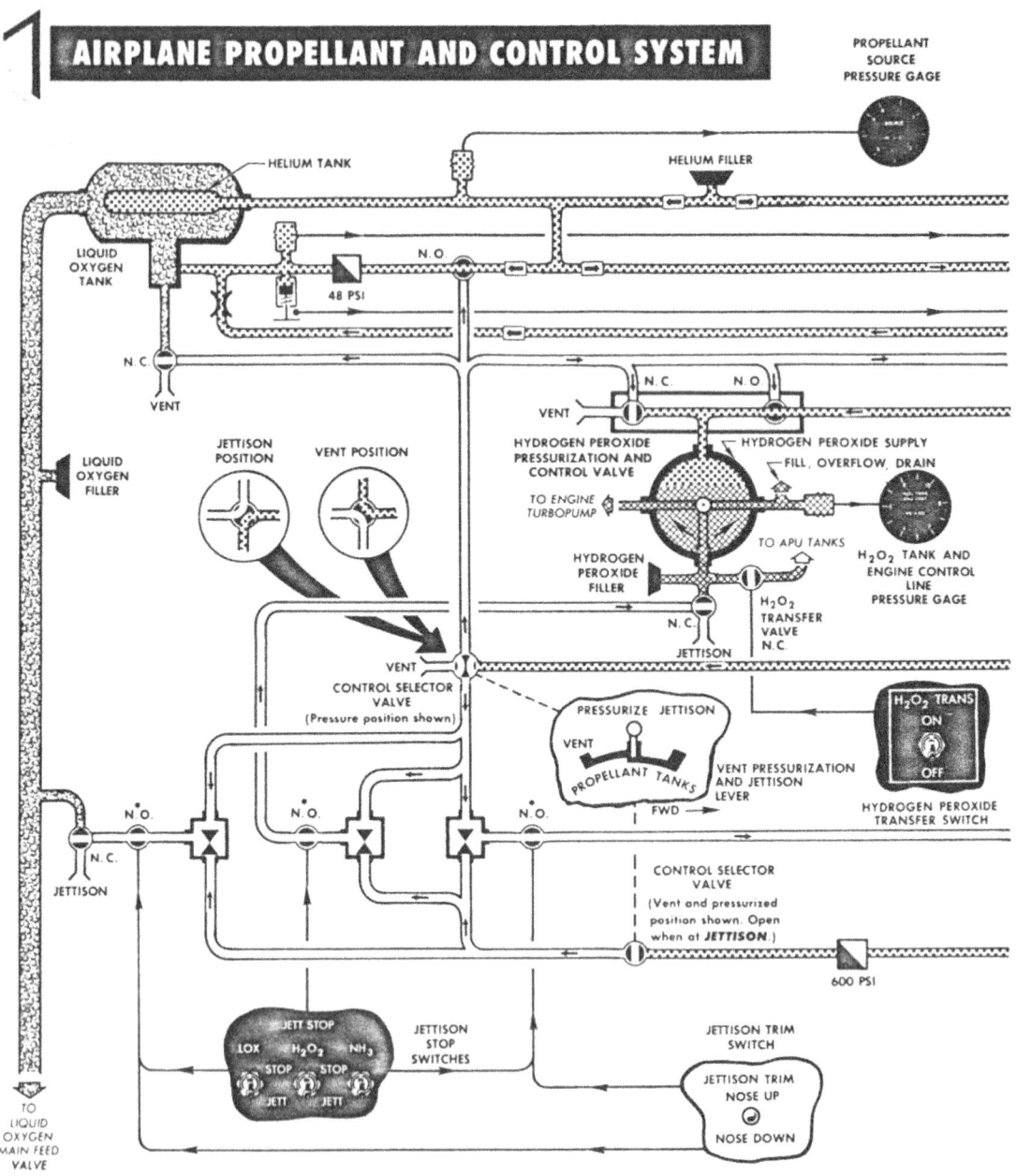

NOTE
Move jettison trim switch to **NOSE DOWN** to stop liquid oxygen jettison. Move switch to **NOSE UP** to stop ammonia jettison.

X-15-1-48-5D

Figure 1-6

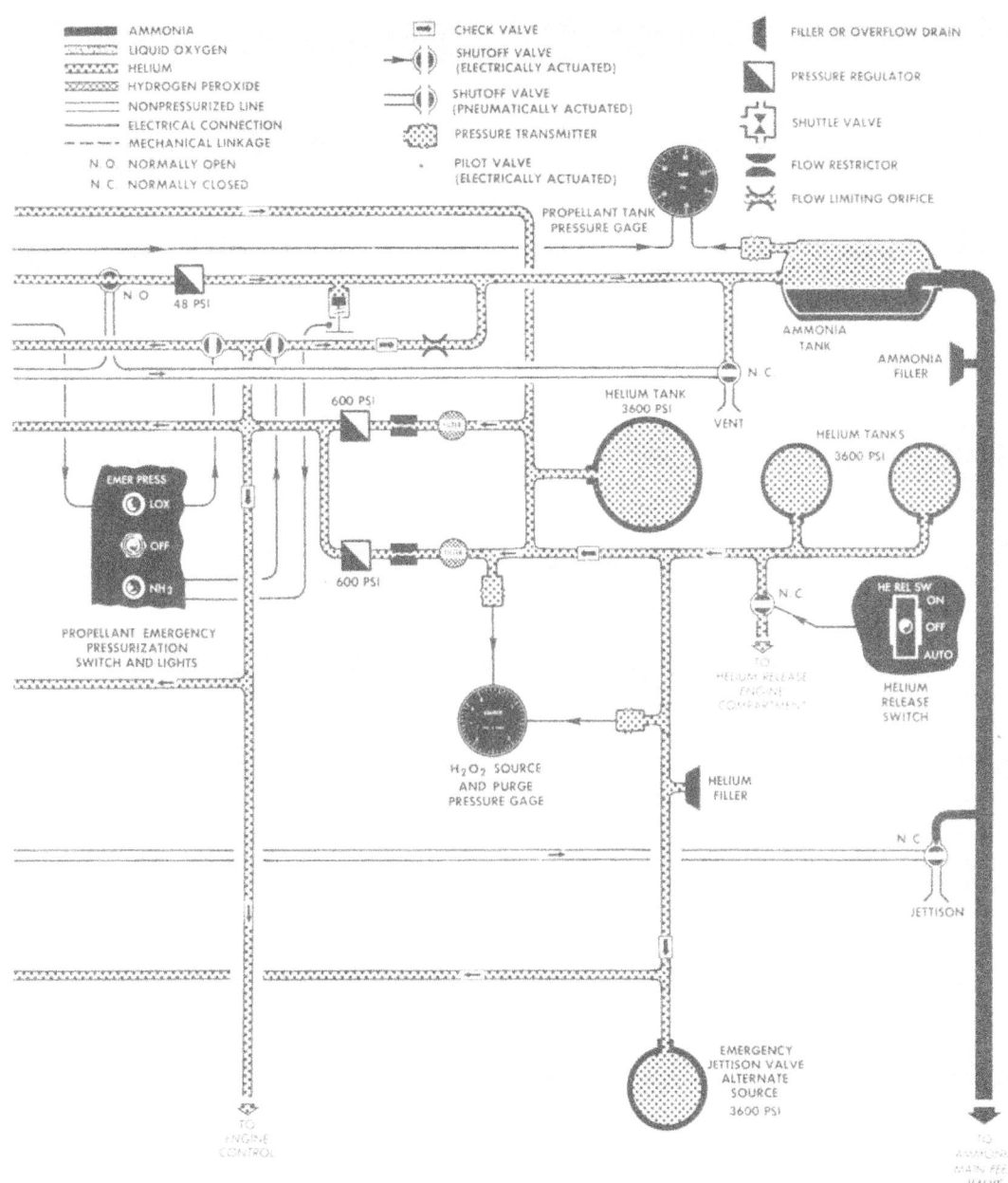

NOTE

When the vent, pressurization, and jettison control lever is in the **VENT** position, all vent valves are open, and pressurization and jettison valves are closed. When the lever is in the **PRESSURIZED** position, the pressurization valves are open, and the jettison and vent valves are closed. When the lever is in the **JETTISON** position, the jettison and pressurization valves are open, and the vent valves are closed.

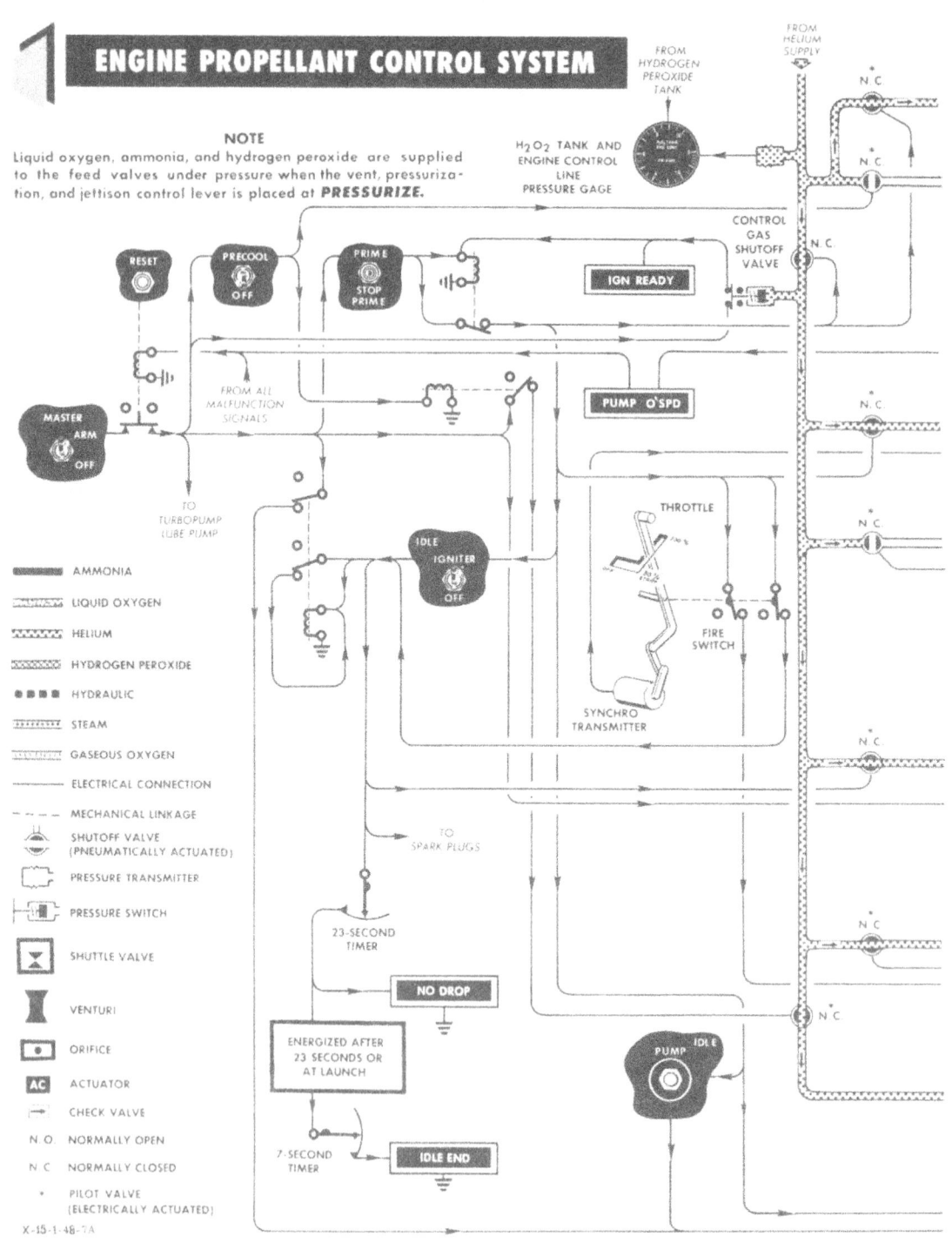

Figure 1-7

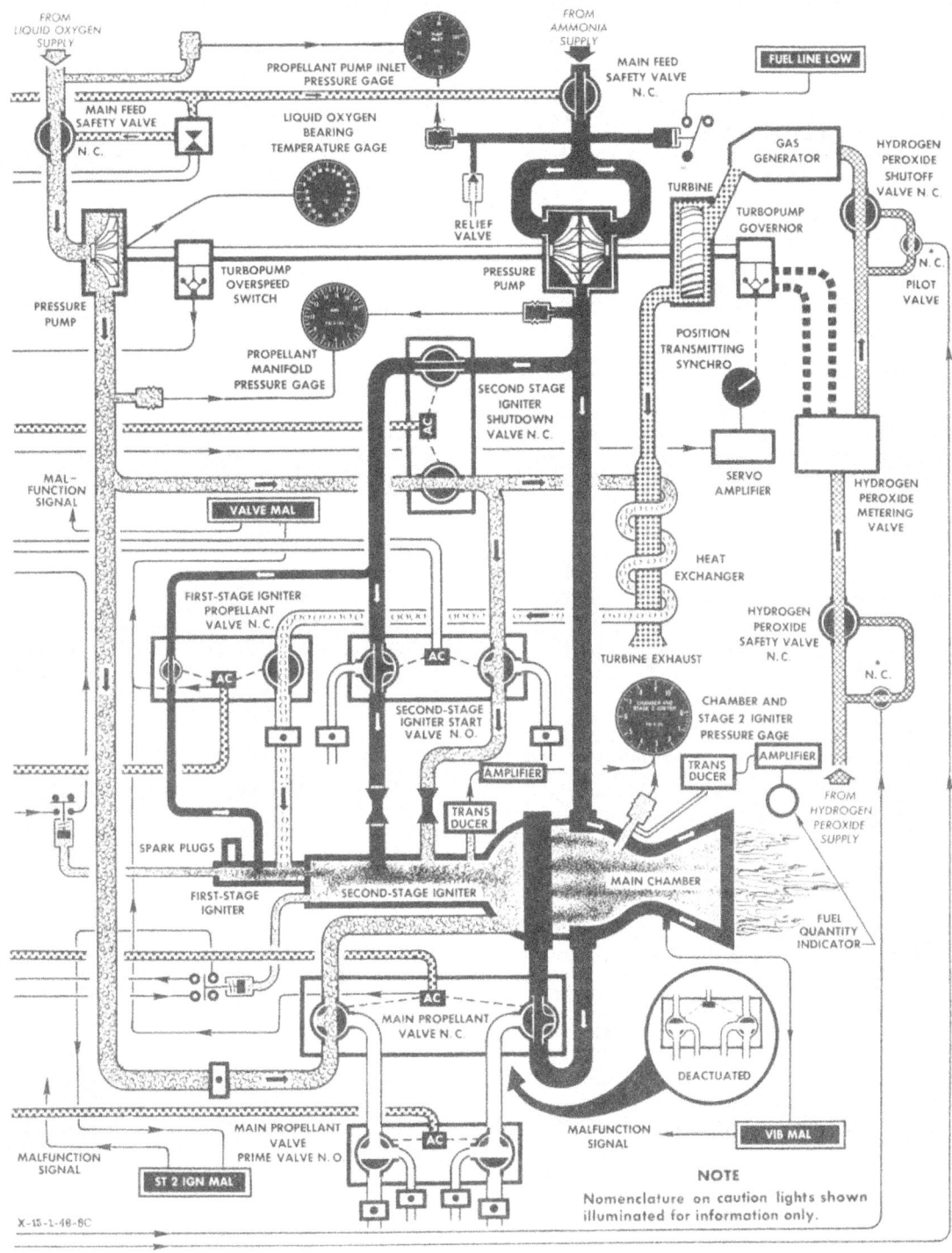

Section I   T.O. 1X-15-1

Turbopump Idle Button.

The turbopump idle button (52, figure 1-2), on the instrument panel, is powered by the primary dc bus through the engine master switch and the engine overspeed reset button. With an engine start sequence initiated and the prime phase completed, depressing this button for one second opens the turbopump H_2O_2 downstream shutoff valve, which starts turbopump operation. When pressure of the ammonia in the propellant manifold builds to approximately 210 psi, the turbopump speed control system begins operation and maintains the turbopump at idle speed.

Igniter Idle Switch.

This switch (49, figure 1-2), on the instrument panel, receives power from the primary dc bus through the engine master switch and engine overspeed reset button. With an engine start sequence initiated and the engine turbopump operating at idle speed, moving this switch from OFF (down) to IGNITER causes the following sequence of actions: a 2-second helium purge is initiated, and the spark plugs are energized; the first-stage igniter propellant valves open, and igniter and idle timing starts; gaseous nitrogen flow (from the carrier airplane) starts when first-stage igniter pressure reaches a specified value; propellants flow to the second-stage igniter; and second-stage ignition occurs.

WARNING

This phase of operation (igniter idle) is limited to 30 seconds. Either igniter idle operation must be terminated (by placing the engine prime switch to STOP PRIME) or the launch accomplished at the end of the 30-second idle period.

ENGINE INDICATORS.

Propellant Tank Pressure Gage.

This dual-indicating gage (63, figure 1-2), on the instrument panel, is powered by the 26-volt ac bus. The gage indicates the two propellant tank pressures. The gage is graduated from 0 to 100 psi in increments of 5 psi. One pointer of the gage has the letter "L," indicating the liquid oxygen tank pressure; the other pointer has the letter "A," indicating the ammonia tank pressure.

Propellant Pump Inlet Pressure Gage.

This dual-indicating gage (60, figure 1-2) is powered by the 26-volt ac bus. It indicates liquid oxygen and ammonia pressures at the engine turbopump inlets. This gage is graduated from 0 to 100 psi in increments of 25 psi. The pointer labeled "L" reads the liquid oxygen pressure; the pointer labeled "A" indicates the ammonia pressure.

Liquid Oxygen Bearing Temperature Gage.

This gage (47, figure 1-2), labeled "LOX BRG TEMP," is powered by the 26 volt ac bus and indicates the temperature of the bearing for the liquid oxygen centrifugal pump segment of the engine turbopump. The gage is graduated from 50°C to 130°C in 10-degree increments.

Propellant Manifold Pressure Gage.

This dual-indicating gage (57, figure 1-2), on the instrument panel, is powered by the 26-volt ac bus. The gage indicates propellant pump discharge pressures. It is graduated from 0 to 2000 psi in increments of 50 psi. One pointer is labeled "L" and indicates liquid oxygen pump discharge pressure; the other pointer is labeled "A" and indicates ammonia pump discharge pressure.

Chamber and Stage 2 Igniter Pressure Gage.

This dual-indicating gage (51, figure 1-2), on the instrument panel, is powered by the 26-volt ac bus. The gage is graduated from 0 to 1000 psi in increments of 20 psi from 0 to 100, and 50 psi from 100 to 1000. The short hand indicates pressure in the second-stage igniter. The long hand indicates pressure in the main thrust chamber.

Ignition-ready Indicator Light.

This green indicator light (1, figure 1-2), on the instrument panel, is powered by the primary dc bus through the engine master switch. When illuminated, it reads "IGN READY," indicating that the engine electrical circuits and purge gas network have been energized. In the normal starting sequence, this light will go out for 2 seconds when the igniter idle switch is moved to IGNITER, then come on again. During all helium purges, this light will go out. This light may be tested by the indicator, caution, and warning light test circuit.

No-drop Caution Light.

This amber caution light (79, figure 1-2) on the instrument panel, is powered by the primary dc bus through the engine master switch. During a normal engine start sequence, this light will come on when 7 seconds remains in the igniter idle phase of operation. The light, when illuminated, reads "NO DROP" and serves to warn the pilot to terminate igniter idle operation or to continue on to the launch phase. This light may be tested by the indicator, caution, and warning light test circuit.

Idle-end Caution Light.

This amber caution light (78, figure 1-2), on the instrument panel, is powered by the primary dc bus through the engine master switch. This light will illuminate, reading "IDLE END," when the 30-second igniter idle phase of engine operation is complete. When this light comes on, engine shutdown must be accomplished or operation continued into the main chamber phase (after launch). This light may be tested by the indicator, caution, and warning light test circuit.

Valve Malfunction Caution Light.

This amber caution light (76, figure 1-2), on the instrument panel, is powered by the primary dc bus through the engine master switch. When illuminated, this light reads "VALVE MAL." This light will come on when

1-14   CHANGED 26 APRIL 1963

malfunction shutdown occurs because the main or first-stage propellant valve is improperly positioned during the starting sequence. The light will also come on momentarily whenever a malfunction shutdown occurs with the main chamber operating. This light may be tested by the indicator, caution, and warning light test circuit.

Stage 2 Ignition Malfunction Caution Light.

This amber caution light (74, figure 1-2), on the instrument panel, is powered by the primary dc bus through the engine master switch. When illuminated, this light reads "ST 2 IGN MAL." The light will come on when a malfunction shutdown occurs during the starting sequence because of failure of the second-stage igniter to reach operating pressure. The light will also come on momentarily whenever a malfunction shutdown occurs with the main chamber operating. This light may be tested by the indicator, caution, and warning light test circuit.

Vibration Malfunction Caution Light.

This amber light (71, figure 1-2), on the instrument panel, comes on when the engine shuts down because of excessive vibration. Excessive engine vibration causes a shutdown signal to be transmitted from either of two sensors to the engine control box. The signal causes the actuation of two malfunction relays in the control box that de-energize the prime, precool, and firing circuits to shut down the engine and turn on the light. The light is powered by the primary dc bus. If the light comes on during powered flight, an engine restart may be attempted.

Turbopump Overspeed Caution Light.

This amber caution light (73, figure 1-2), on the instrument panel, is powered by the primary dc bus through the engine master switch. When illuminated, this light reads "PUMP O'SPD." The light will come on if a malfunction shutdown occurs because of turbopump overspeed which is not corrected by the turbopump governor. This light may be tested by the indicator, caution, and warning light test circuit.

(Deleted)

Fuel Line Low Caution Light.

An amber "FUEL LINE LOW" caution light (56, figure 1-2) is on the left side of the instrument panel. This light, powered by the primary dc bus through the engine master switch, is actuated by a pressure switch installed in the fuel (ammonia) line downstream of the main safety valve. If fuel pressure at the turbopump inlet drops to 32 (±2) psi, the light will come on. Illumination of the light indicates that partial cavitation of the pump is likely to occur.

CAUTION

When the "FUEL LINE LOW" caution light is on, there is not sufficient cooling around the main chamber of the engine for temperature protection. Thrust settings above 59% (40%) may increase the temperature and cause damage to the engine.

If the light comes on before the engine is started, the start will be aborted. (Refer to "Fuel Line Pressure Low" in Section III.) When this light has been illuminated, it will remain on until the engine master switch has been placed in the OFF position. The light may be tested by the indicator, caution, and warning light test switch.

PROPELLANT SUPPLY SYSTEM.

The propellant supply system consists of the liquid oxygen supply (oxidizer), anhydrous ammonia (fuel) supply, valves, and associated plumbing. The liquid oxygen and ammonia are fed under a low inert gas pressure from the supply tanks to the turbopump for engine operation. The helium supply systems, which furnish gas pressure for tank pressurization, pneumatic valve operation, and system purging, are described in separate paragraphs in this section. See figure 1-16 for the liquid oxygen and ammonia specifications.

LIQUID OXYGEN TANK.

The liquid oxygen supply is carried in a triple-compartmented tank (12, figure 1-1), just aft of the No. 2 equipment compartment. The center section area of the cylindrical tank is hollow and forms a case for a gaseous helium high-pressure storage tank. When the liquid oxygen tank is not under pressurization, it is vented to atmosphere. The tank compartments are check-valve-vented. Each compartment feeds rearward toward the airplane center of gravity. The liquid oxygen is fed from the rear compartment under 48 psi of helium pressure to the turbopump or jettison line through a series of control valves. The total volume of the tank is 1034 US gallons; of this amount, 14 gallons is residual at a liquid surface angle of 38 degrees, and 17 gallons is vent and expansion space. The total usable liquid oxygen is 1003 gallons. The tank is filled for flight through the carrier airplane's supply system. The tank incorporates a liquid oxygen fluid level sensing switch that permits the tank to be topped off automatically whenever fluid drops below a predetermined level. For ground operational checks, the tank is serviced through the receptacle mounted on the engine feed line. The tank filler is on the topside of the wing fairing tunnel forward of the left wing root leading edge.

AMMONIA TANK.

The ammonia supply is carried in a triple-compartmented cylindrical tank (17, figure 1-1), just aft of the No. 3 equipment compartment and ahead of the turbopump hydrogen peroxide tank. The center section area is hollow and closed at both ends. The rear compartment center section is perforated to allow storage of ammonia within the center section area. The compartments are check-valve-vented. This aids in the pressure feed of the fluid transfer from the rear tank compartment forward toward the airplane center of gravity. The rear compartment empties first; then the middle compartment empties into the front compartment, with the ammonia fed from the front compartment under 48 psi of helium pressure to the turbopump or jettison line through a series of control valves. The total volume of the tank is approximately 1445 US gallons. The tank is ground-serviced only. The filler receptacle for the tank is on the underside of the right wing root fairing tunnel.

ENGINE AND PROPELLANT CONTROL HELIUM SYSTEM.

Helium to pressurize the turbopump hydrogen peroxide supply tank and to supply pneumatic pressure for engine and propellant control is contained in four spherical tanks. One tank (15, figure 1-1) is between the liquid oxygen and ammonia tanks. Two tanks (23, figure 1-1) are in the left and right wing root fairing tunnels outboard of the engine. These three tanks are interconnected, supplying 3600 psig pressure to two pressure-reducing regulators in parallel. The fourth tank is just to the right of the turbopump H_2O_2 supply tank and supplies helium at 3600 psig to a single pressure-reducing regulator for emergency or secondary pneumatic control of the propellant jettison valves. This tank is interconnected with the other three tanks for filling purposes only. From the parallel pressure-reducing regulators of the main supply, helium at 575 to 600 psig is supplied to the engine helium manifold for operation of engine control valves, to the turbopump H_2O_2 supply tank for tank pressurization, and to propellant control jettison and main feed valves. Two of the tanks supply helium directly to the helium dump valve, for engine compartment purging. The dump valve is solenoid-operated and controlled by the helium release selector switch. For information on operation of this switch, refer to "Helium Release Selector Switch" in this section. The helium to the engine helium manifold is in turn routed to a control gas valve and the two gas regulators in the purge valve network at a pressure of 550 to 600 psig. The control gas valve is energized during the prime period and admits helium at a pressure of 550 to 600 psig to the pilot valves for the prime valve, first-stage igniter start valve, second-stage igniter start and shutoff valves, and main propellant valve. Helium at a pressure of 125 to 200 psig is routed from the two purge gas regulators and to the return side of the second-stage igniter start and shutoff valves. Another regulator supplies helium from the helium manifold at 7.5 psig to the lubrication system accumulator, engine control box, and hydraulic power package.

HELIUM RELEASE SELECTOR SWITCH.

Refer to "Engine Compartment Purging System" in this section.

H_2O_2 SOURCE AND PURGE PRESSURE GAGE.

A dual-indicating H_2O_2 source and purge pressure gage (64, figure 1-2), on the instrument panel, is powered by the 26-volt ac bus. This gage indicates the helium pressure available from three of the engine and propellant helium system tanks. Needle 1 indicates pressure in the large tank between the liquid oxygen and ammonia tanks. Needle 2 indicates pressure in the two smaller tanks in the wing root fairing tunnels. The gage is calibrated from 0 to 4000 psi in increments of 100 psi. Normally, the two pointers will indicate the same pressure. However, if there is a malfunction of the emergency jettison system helium supply or if helium is dumped into the engine compartment, the pointers will not indicate the same pressure. There is no gage in the cockpit which indicates pressure in the emergency jettison system helium supply tank.

H_2O_2 TANK AND ENGINE CONTROL LINE PRESSURE GAGE.

This dual-indicating gage (61, figure 1-2), on the instrument panel, is powered by the 26-volt ac bus. One pointer, labeled "C," indicates engine control line (helium) pressure downstream of the two parallel pressure regulators. The other pointer, labeled "T," indicates pressure in the turbopump H_2O_2 supply tank. The gage is calibrated from 0 to 1000 psi in increments of 50 psi, except that the range 0 to 100 is in increments of 20 psi.

PROPELLANT PRESSURIZATION HELIUM SYSTEM.

The propellant pressurization helium system supplies gas to pressurize the liquid oxygen and ammonia tanks. This helium is contained in the supply tank (14, figure 1-1) within the center section of the liquid oxygen tank and is pressurized to 3600 psi. The helium flows to the normally open pressure regulators of the liquid oxygen and ammonia supply tanks. The two regulators are actuated by helium pressure (from the engine and propellant helium control system) to the closed position when the vent, pressurization, and jettison control lever is at VENT. When the control lever is placed at PRESSURIZE or JETTISON, the regulators open and helium pressure flows to the liquid oxygen and ammonia tanks.

The regulators reduce the helium pressure to 48 psi before it enters the liquid oxygen and ammonia tanks. When the liquid oxygen and ammonia tanks are pressurized, the propellants are forced through the feed lines to the main feed shutoff valves.

PROPELLANT SOURCE PRESSURE GAGE.

The propellant source pressure gage (62, figure 1-2), on the instrument panel, is powered by the 26-volt ac bus. The gage indicates pressure in the cylindrical helium tank for liquid oxygen and ammonia tank pressurization. The gage is calibrated from 0 to 4000 psi in increments of 100 psi.

PROPELLANT EMERGENCY PRESSURIZATION SYSTEM.

The propellant emergency pressurization system can be used to pressurize either the liquid oxygen or the ammonia tank in case of a failure in the normal pressurization system. This will permit continued low thrust engine operation or propellant jettisoning. The emergency system can supply pressurizing gas to only one propellant tank at a time. The emergency system uses helium from the three interconnected tanks in the engine and propellant control helium system. The system includes a switch and two caution lights.

Propellant Emergency Pressurization Switch.

This three-position switch (75, figure 1-2), on the instrument panel, controls primary dc bus power to the two emergency pressurization system solenoid-operated shut-off valves. With the switch at OFF, the valves are de-energized closed. The switch must be pulled straight out of a detent to move it from OFF to either of the other positions. With the switch at LOX, electrical power is applied to open the shutoff valve which controls emergency helium pressure to the liquid oxygen tank. With the switch at NH_3, electrical power is applied to open the shutoff valve which controls emergency helium pressure to the ammonia tank. All three switch positions are maintained.

Liquid Oxygen and Ammonia Tank Pressure-low Caution Lights.

These lights (72 and 77, figure 1-2), on the instrument panel, are powered by the primary dc bus. The liquid oxygen tank pressure-low caution light is labeled "LOX." The ammonia tank pressure-low caution light is labeled "NH_3." (The nomenclature for the lights also serves as position nomenclature for the propellant emergency pressurization switch.) After the vent, pressurization, and jettison lever is placed at PRESSURIZE, the related light will come on when pressure in the affected tank drops to 34 (±2) psi. If a light comes on during powered flight, it may remain on even after emergency pressurization of the affected tank has been initiated, indicating that the affected tank pressure is not above 40 psi.

NOTE

During the transitional period when the vent, pressurization, and jettison lever is moved from VENT to PRESSURIZE, the lights should come on and remain on for approximately 6 seconds (during build-up of pressure in the propellant tanks).

AUXILIARY POWER UNITS.

The airplane is equipped with two auxiliary power units (7, figure 1-1) that are set side-by-side in a compartment in the forward fuselage. Each unit is a completely automatic, constant-speed, turbine drive machine that transmits power to, and provides structural support for, an ac generator and a hydraulic pump. Propellant for each auxiliary power unit is provided by an independent feed system, using helium pressure to move the monopropellant, hydrogen peroxide. The two auxiliary power units with their respective feed systems are identified as system No. 1 and system No. 2. Their operation is completely independent of each other, and each furnishes one half of the power required. If one unit should fail, the other will provide sufficient electrical and hydraulic power for limited flight capabilities. Each auxiliary power unit is started and stopped by a switch in the cockpit. When an APU is turned on, a solenoid-type shutoff valve is opened to allow hydrogen peroxide from the propellant feed system to flow into the unit. The propellant is routed first through a gear case for cooling purposes (nitrogen gas is also introduced into the upper turbine bearing area for additional cooling) and then to a modulating flow control valve. The flow control valve is modulated to open or close to provide stabilization through a speed control system consisting of a tachometer generator and a frequency detector. Any turbine overspeed condition is sensed by an overspeed sensing element in the speed control system which will automatically act to close the shutoff valve. When the shutoff valve is closed, fuel flow stops and the unit shuts down. The APU shutoff valve is fitted with a drain that opens when the valve is closed, to relieve any excess pressure in the line downstream of the shutoff valve. After passing through the flow control valve, the hydrogen peroxide enters a decomposition chamber containing a catalyst bed. This catalyst bed is made up of a series of silver and stainless-steel screens which act to decompose the hydrogen peroxide into a high-pressure gas mixture of superheated steam and oxygen. The decomposition chamber is heated electrically from the carrier airplane to ensure a fast start under "cold-soak" conditions in case of an emergency. The superheated steam and oxygen mixture enters a nozzle box in the turbine housing. Here, five nozzles convert pressure energy of the fluid into kinetic energy and direct the flow of gas against a turbine wheel. The turbine, acting through a reduction gear train, transmits power to the ac generator and hydraulic pump. The turbine wheel is housed within an exhaust casing which is designed to contain any buckets that might separate from the wheel during an overspeed operation. The exhaust casing collects spent gases that have passed through the turbine wheel and exhausts them overboard. A gear casing assembly contains the reduction gearing, accessory drive pads, cooling passages, provisions for lubrication, and a drive for the tachometer generator. Modified airplanes have a pressurized water cooling system for alternate cooling of the upper turbine bearing. A typical auxiliary power unit and its propellant feed system are shown schematically in figure 1-8. For information on nitrogen and water cooling of the upper turbine bearing of each APU, refer to "APU Cooling Switch" in Section IV.

APU SPEED CONTROL.

The speed control for each auxiliary power unit provides positive speed control, starting and stopping, and

CHANGED 29 JUNE 1962

Section I T.O. 1X-15-1

AUXILIARY POWER UNIT AND BALLISTIC

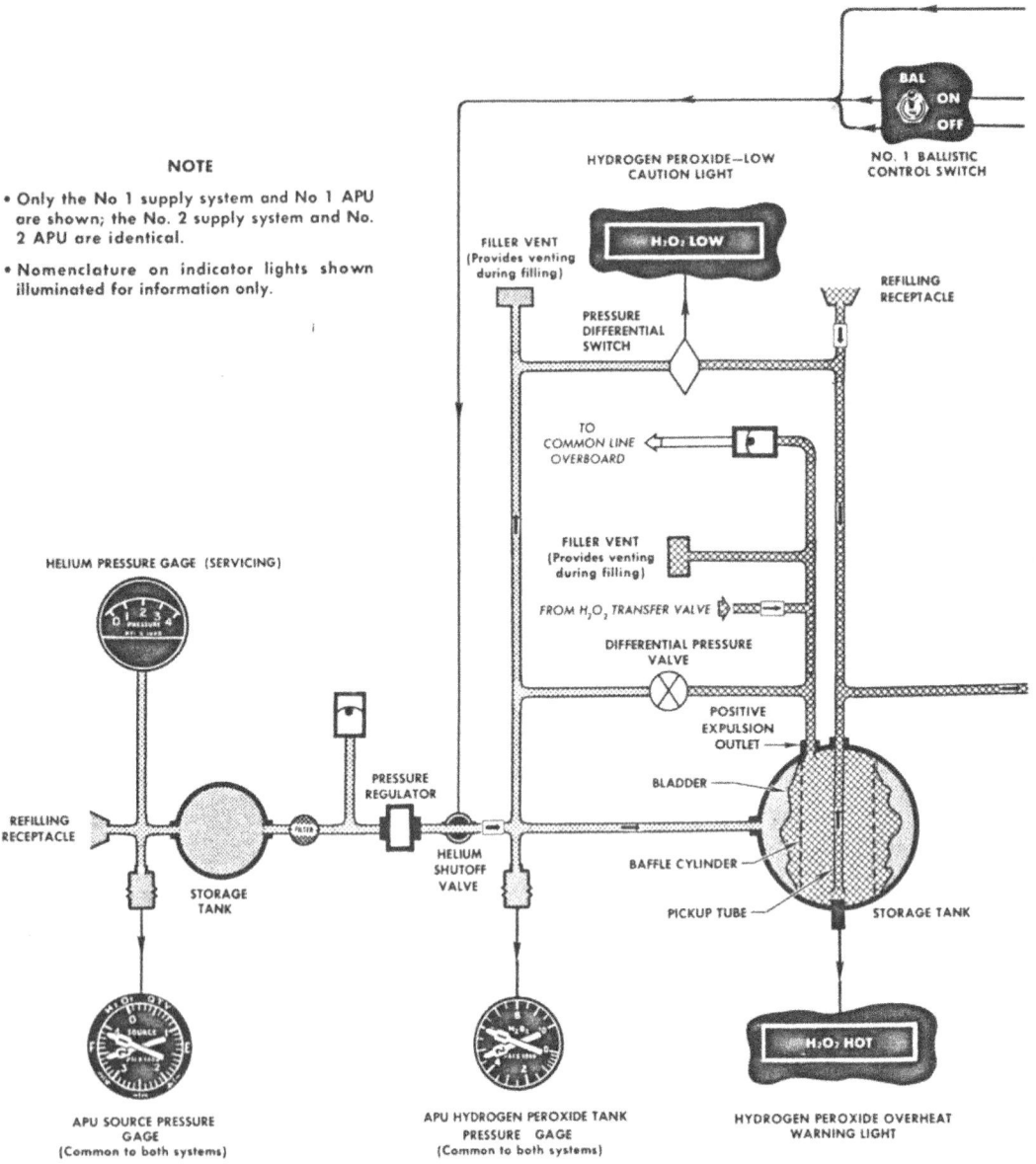

Figure 1-8 (Sheet 1 of 3)

1-18 CHANGED 29 JUNE 1962

CONTROL AND REACTION AUGMENTATION SYSTEMS

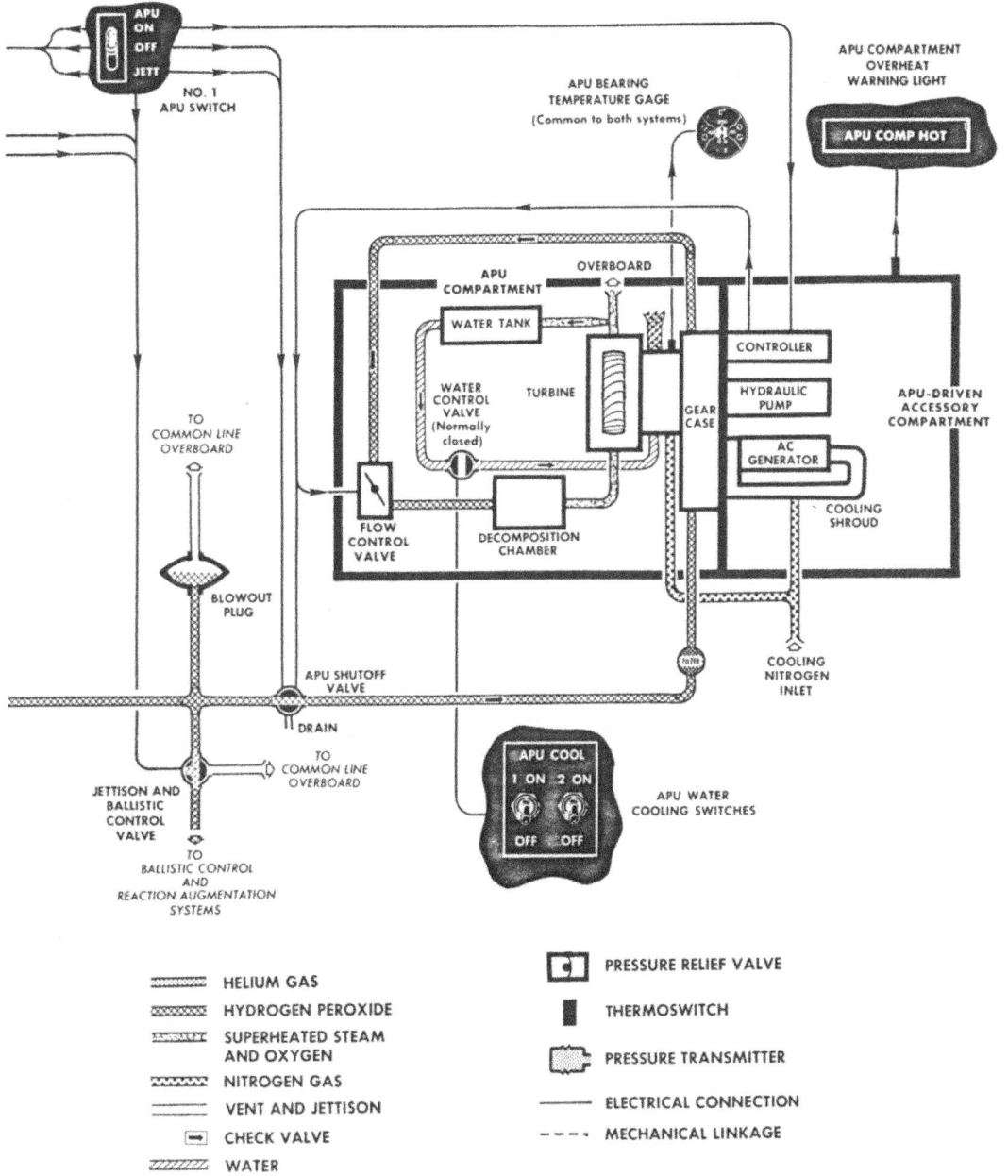

Figure 1-8 (Sheet 2 of 3)

AUXILIARY POWER UNIT AND BALLISTIC CONTROL SYSTEMS

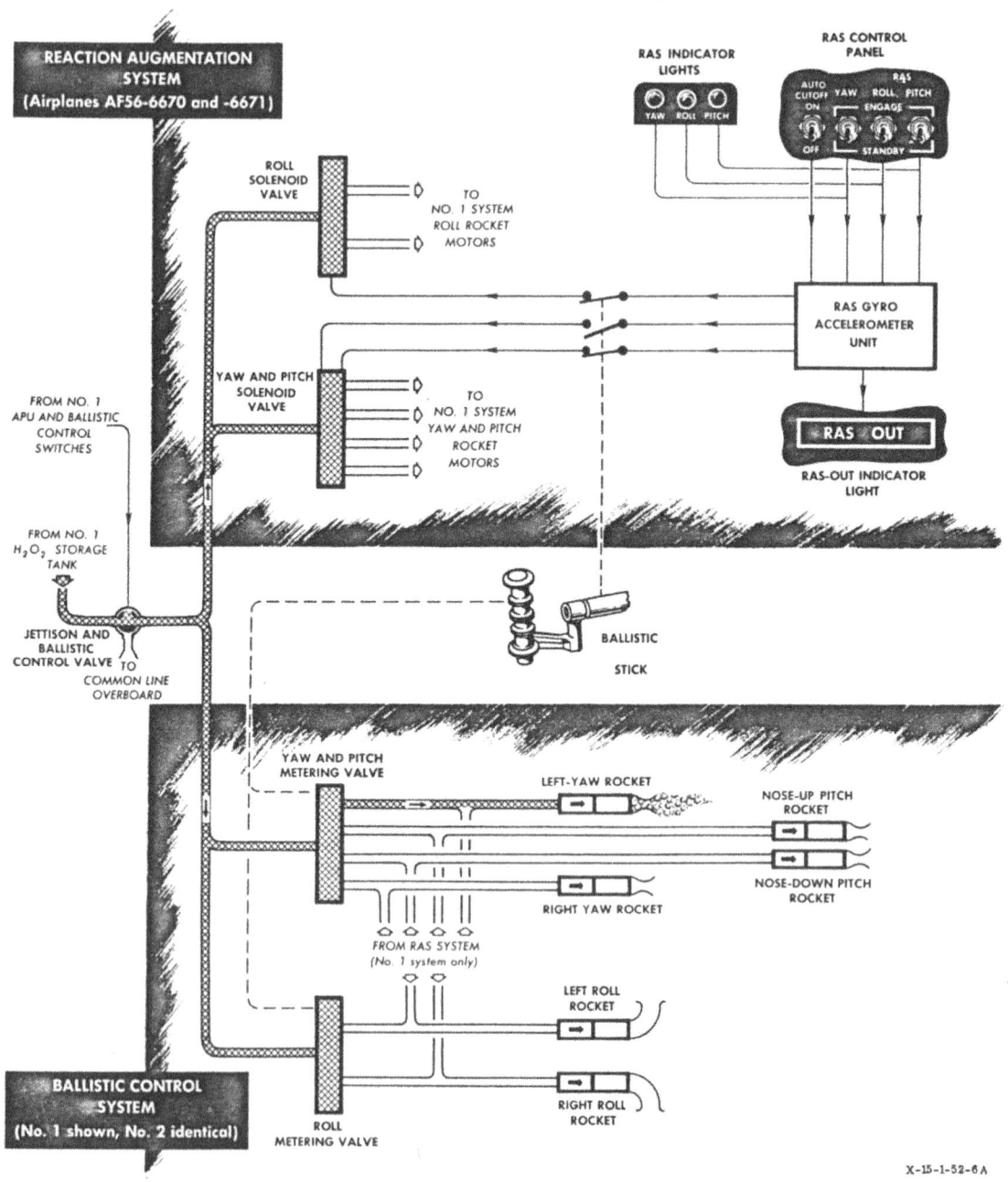

Figure 1-8 (Sheet 3 of 3)

overspeed protection by regulating the flow of hydrogen peroxide to the decomposition chamber. The key component of each speed control is a controller which contains the necessary circuits for sensing unit operation through the frequency output of a tachometer generator. During normal operation, the frequency of the power generated by the tachometer generator, being proportional to the turbine speed, matches a preset frequency of the controller. Any change in turbine speed due to a change in load, or from any other disturbance, causes a proportional change in frequency of the tachometer generator. This frequency change is sensed by the controller, which in turn adjusts the opening of the flow control valve to bring turbine speed back to the normal operating level. During normal operation, the speed of the unit is automatically controlled to maintain 51,200 rpm by the speed-sensing element of the controller. Should an overspeed condition occur (56,000 rpm or greater), the overspeed sensing element of the controller automatically acts to energize the solenoid-operated APU shutoff valve to the closed position, thereby shutting off the flow of the propellant. The unit then decelerates and stops. It cannot be restarted until the APU switch is first cycled to the OFF position.

APU SWITCHES.

There are two APU switches (23 and 46, figure 1-2) on the instrument panel, one for control of each auxiliary power unit and its associated feed system. When either switch is turned to ON, battery-bus power is used to open the helium shutoff valve in the related propellant feed system, allowing helium pressure to move the hydrogen peroxide through the feed lines. At the same time, power is applied to the opening circuit of the related APU shutoff valve. This permits the propellant to flow to the auxiliary power unit. Turning the switch to OFF closes the helium shutoff valve (if the ballistic control switch is at OFF), shutting off the helium supply.

NOTE

If the ballistic control switch is at ON, the OFF position of the APU switch will not close the helium shutoff valve.

At the same time, the APU shutoff valve closes, shutting off the flow of the propellant to the unit. The JETT position, powered by the primary dc bus, is used if an emergency arises and it is desired to jettison the propellant overboard. The switch is guarded to prevent it from being accidentally moved to the JETT position. When the switch is turned to JETT, the following occurs: The helium shutoff valve in the feed system opens, or remains open if it is already so, allowing the helium to continue to force the hydrogen peroxide through the feed lines. Concurrently, the APU shutoff valve closes and a jettison and ballistic control valve in the feed system opens to the jettison position. The jettison and ballistic control valve serves both as a shutoff valve for propellant supply to the ballistic control system and as a propellant jettison control. As this valve opens to the jettison port, the propellant is routed through a line that dumps overboard at the aft end of the airplane.

APU COMPARTMENT OVERHEAT CAUTION LIGHTS.

Two APU compartment amber overheat caution lights (21 and 40, figure 1-2) are adjacent to their related APU switches on the instrument panel. The lights are powered by the battery bus. A thermoswitch in each APU accessory drive compartment is set to energize the light when the temperature in the compartment rises to approximately 525° F. The lights read "APU COMP HOT" when on. If either light comes on, the related auxiliary power unit should be shut down immediately.

APU AND BALLISTIC CONTROL PROPELLANT FEED SYSTEMS.

Two completely independent feed systems provide propellant for the auxiliary power units and the ballistic control system. (Refer to "Ballistic Control System" in this section.) System 1 is in the left side of the fuselage; system 2 is in the right side of the fuselage. The systems are identical. Helium gas under pressure moves the monopropellant hydrogen peroxide to its basic function of providing fuel to these units at the required flow rates and pressures. Each propellant feed system includes a high-pressure, spherical, helium storage tank and a positive expulsion-type hydrogen peroxide storage tank. Both tanks are below the related auxiliary power unit in the forward fuselage. Helium and hydrogen peroxide filler valves and helium high-pressure gages for ground servicing are in each side of the fuselage side fairings. Helium and hydrogen peroxide pressure gages common to both systems are in the cockpit. When the ballistic control switch is turned ON, or when the APU switch is turned ON (or to JETT), a shutoff valve is opened to allow helium to flow from the storage tank. The helium tank contains enough helium to expel all the hydrogen peroxide in the hydrogen peroxide storage tank. Helium pressure is reduced from 3600 psi at the tank to 550 psi as it passes through a pressure regulator. A relief valve upstream of the pressure regulator prevents overpressurization due to overcharging or pressure build-up during high-temperature conditions. From the pressure regulator, the helium passes through the shutoff valve and pressurizes the hydrogen peroxide tank. The positive-expulsion type hydrogen peroxide tank contains a baffle cylinder, perforated to allow the propellant to flow to a pickup tube inside the baffle cylinder. The inlet of the pickup tube is very close to the bottom of the tank to prevent it from being uncovered during normal flight attitudes when only approximately 20 percent of the propellant supply remains in the tank. Between the baffle cylinder and tank wall is a collapsible plastic bladder. The helium enters the tank between the wall and the bladder where pressure on the bladder forces the hydrogen peroxide into the baffle cylinder through the pickup tube and out of the tank. A check valve upstream of the tank prevents hydrogen peroxide from backing into the helium system in case of a bladder failure. When the tank is emptied to the extent that the bladder collapses against the baffle cylinder, the feed pressure will drop off. This pressure drop creates a pressure differential between the helium and hydrogen peroxide. When this pressure differential increases to approximately 35 psi, a differential pressure switch in the system actuates a low-level caution light in the cockpit. Pressure differential is also sensed

Section I   T.O. 1X-15-1

by a valve that opens at approximately 55 psi and allows helium to flow to the top of the baffle cylinder, expelling the remaining hydrogen peroxide. Protection of the hydrogen peroxide tank against rupture due to overpressurization is provided by a pressure relief valve and a blowout plug. The relief valve is designed to open at approximately 650 psi (100 psi over normal tank pressure). In case of a malfunction of the pressure relief valve or an abnormal rate of pressure increase, the blowout plug will rupture at approximately 900 psi. If the blowout plug should rupture, the affected system will be deactivated by loss of the propellant through the plug. The flow of both of these relief devices is routed overboard through a vent and jettison line. A thermoswitch at the base of the tank energizes a warning light in the cockpit if the hydrogen peroxide in the tank becomes overheated. From the storage tank, the propellant is routed into feed lines through shutoff valves to the auxiliary power unit and the ballistic control system. Temperature of the propellant at the APU inlet must be a minimum of 40° F during starting. To prevent freezing in the feed lines, warm air from the carrier airplane is pumped into the compartment containing the propellant feed system hydrogen peroxide components to maintain a temperature of approximately 120° F. The system is designed to dump helium and hydrogen peroxide overboard if an emergency arises.

APU and Ballistic Control Propellant Feed System Controls and Indicators.

APU Source Pressure Gage. A dual-movement helium pressure gage (42, figure 1-2), common to both propellant feed systems, is on the right side of the instrument panel. The gage, labeled "SOURCE," is calibrated in pounds per square inch and has two pointers, marked "1" and "2" for system identification. The gage includes a slip ring, with calibration markings of "F," "3/4," "1/2," "1/4," and "E." The slip ring is used to indicate the amount of H_2O_2 available to the APU's. The quantity of H_2O_2 available from the tanks is proportional to source (helium) pressure. Therefore, the slip ring should be rotated just before APU start so that the "F" mark is aligned with the No. 1 pointer; then the position of the No. 1 pointer in relation to the slip ring calibrations will indicate the amount of H_2O_2 available for APU operation once they are started. The pressure indicating system is powered by the 26-volt ac bus. Operating pressure will vary from 3600 down to 550 psi, depending upon the helium supply in the storage tank.

APU Hydrogen Peroxide Tank Pressure Gage. A dual-movement hydrogen peroxide pressure gage (37, figure 1-2), common to both propellant feed systems, is on the right side of the instrument panel. The gage is labeled "H_2O_2" and is calibrated in pounds per square inch. The pointers are marked "1" and "2" for system identification. The gage shows tank pressure in the hydrogen peroxide storage tanks, sensed by a pressure transmitter in each feed system. Normal operating pressure is approximately 550 psi. The pressure indicating system is powered by the 26-volt ac bus.

APU Bearing Temperature Gage. A dual-pointer APU bearing temperature gage (38, figure 1-2) is on the instrument panel. The gage shows in degrees centigrade the temperature of No. 1 and No. 2 APU upper turbine bearings. The temperature indicating system is powered by the No. 1 primary ac bus. The gage is calibrated from zero to 200 in increments of 20 degrees. The left pointer indicates No. 1 APU upper turbine bearing temperature; the right pointer, No. 2 APU upper turbine bearing temperature.

Hydrogen Peroxide-low Caution Lights. Each propellant feed system has a low-level caution light (22 and 43, figure 1-2) that reads "H_2O_2 LOW" when on. The two amber, placard-type lights, on the right side of the instrument panel, are powered by the primary dc bus. A pressure-differential switch in each system becomes energized, causing the related light to come on when the differential pressure between helium and hydrogen peroxide rises to approximately 35 psi. The light will come on at this instant, when approximately 20 percent of the hydrogen peroxide supply is left in the storage tank. If either light comes on, extreme maneuvers should be avoided to prevent uncovering the inlet of the pickup tube in the tank, thus allowing helium to flow into the hydrogen peroxide line.

NOTE

During hydrogen peroxide transfer, the hydrogen peroxide low caution light will go out whenever the pressure differential drops below approximately 35 psi.

Hydrogen Peroxide Overheat Warning Lights. The No. 1 and No. 2 hydrogen peroxide overheat warning lights (20 and 39, figure 1-2) are on the right side of the instrument panel. The red, placard-type lights are powered by the battery bus and read "H_2O_2 HOT" when on. A thermoswitch at the base of each system hydrogen peroxide storage tank energizes the related light if the temperature of the contents of the tank rises to approximately 160°F. If either light comes on, the contents of the affected tank should be jettisoned. Concurrently, the related auxiliary power unit will automatically shut down.

APU Switches. Refer to "Auxiliary Power Units" in this section.

Ballistic Control System Switches. Refer to "Ballistic Control System" in this section.

HYDROGEN PEROXIDE TRANSFER SYSTEM.

Modified airplanes have a transfer system incorporated in the hydrogen peroxide propellant feed systems to provide a means of transferring hydrogen peroxide from the engine turbopump propellant tank (figure 1-6) to the APU and ballistic control propellant tanks (figure 1-8) after engine shutdown. Transfer may be accomplished whenever one or both hydrogen peroxide low caution lights come on.

Hydrogen Peroxide Transfer Switch.

The hydrogen peroxide transfer switch (16A, figure 1-2), labeled "H_2O_2 TRANS," is on the upper right side of the instrument panel. It has three positions: ON, OFF, and an unmarked center neutral position. When the

1-22   CHANGED 29 JUNE 1962 switch is at ON, primary dc bus power opens a solenoid-operated transfer valve and allows flow of hydrogen peroxide from the engine turbopump tank to the APU tanks. The switch will return to the spring-loaded center position when released, leaving the valve in the desired position. To stop transfer of hydrogen peroxide, the switch is placed at OFF.

ELECTRICAL POWER SUPPLY SYSTEMS.

The airplane is equipped with an alternating-current and a direct-current electrical power system. (See figure 1-9.) Power for the ac system is supplied by two alternator-type generators. The dc system normally is powered from the ac system through two transformer-rectifiers. A 24-volt battery is available for use in an emergency to supply dc power to essential equipment. During ground operation, ac and dc power can be supplied to the airplane by an external power source. During captive flight, the carrier airplane can supply ac and dc power to the airplane. Both external power sources supply minimum dc power for initial relay or valve operation only. Large amounts of dc power then are supplied from the ac system through the transformer-rectifiers.

AC ELECTRICAL POWER DISTRIBUTION.

Two ac generators supply 200/115-volt, 400-cycle, three-phase ac power to the two primary ac busses. Each generator is driven through a gear train by an auxiliary power unit. (Refer to "Auxiliary Power Units" in this section.) Two 26-volt ac busses are powered by the No. 2 primary ac bus. Automatic frequency control and voltage regulation are provided for the ac generators.

THIS PAGE INTENTIONALLY LEFT BLANK.

AC Generators.

Each ac generator is driven through a gear train by an auxiliary power unit and supplies 200/115-volt, 400-cycle, three-phase ac power to its respective primary ac bus. Failure of an APU causes failure of the ac generator it drives. If one ac generator fails for any reason, the other ac generator automatically supplies power to both primary ac busses. If either generator drops "off the line" because of a momentary malfunction, it can be reset "onto the line."

Primary AC Busses.

The No. 1 primary ac bus normally is powered by the No. 1 ac generator; the No. 2 primary ac bus, by the No. 2 ac generator. However, if either generator fails, the remaining generator will power both primary ac busses. External power, on the ground or from the carrier airplane, will power the primary ac busses, but only when neither ac generator is on.

26-volt AC Busses.

The two 26-volt ac busses are powered through two parallel transformers by the No. 2 primary ac bus. The 26-volt busses are powered as long as either ac generator is operating. In addition, when external power is applied to the airplane on the ground or from the carrier airplane (and both ac generators are off), the 26-volt busses are powered.

DC ELECTRICAL POWER DISTRIBUTION.

Direct-current power is distributed from the 28-volt primary dc bus and the battery bus.

28-volt Primary DC Bus.

The 28-volt primary dc bus is powered by both primary ac busses through two transformer-rectifiers. The primary dc bus, in addition to powering certain equipment, normally powers the battery bus. Failure of one ac generator will not de-energize the primary dc bus. The primary dc bus also is energized when external power is applied on the ground or from the carrier airplane.

Battery Bus.

The battery bus normally is powered by the primary dc bus. However, the emergency battery can be connected to the battery bus to provide emergency dc power. In addition, external power on the ground or from the carrier airplane can be applied to the battery bus.

Emergency Battery.

A stand-by, 24-volt, emergency battery is available to provide emergency power to the battery bus.

ELECTRICALLY OPERATED EQUIPMENT.

See figure 1-9.

EXTERNAL ELECTRICAL POWER RECEPTACLE.

The external power receptacle is on the upper surface of the fuselage, aft of the canopy. (See figure 1-16.) When ac and dc external power is applied to the airplane by a ground unit, an adapter must be used. When external power is applied from the carrier airplane, a single plug-in unit in the carrier airplane pylon is used.

CIRCUIT BREAKERS AND FUSES.

The electrical distribution circuits are protected by circuit breakers and fuses on the electrical power panel in the No. 2 equipment compartment and on the circuit-breaker panel (6, figure 1-4) on the right console in the cockpit. All of the circuit breakers on the right console panel are of the push-pull type. The circuit breakers in the No. 2 equipment compartment must be properly positioned before carrier take-off, because they are not accessible in flight.

If the two external power circuit breakers in the No. 2 equipment compartment are not closed before carrier take-off, carrier airplane electrical power cannot be applied to the X-15 Airplane.

ELECTRICAL POWER SUPPLY CONTROLS.

No. 1 Generator Switch.

A three-position switch (15, figure 1-2) on the instrument panel controls operation of the No. 1 ac generator by means of battery bus power. The switch is spring-loaded from the RESET position to ON. When the switch is OFF, the generator is taken "off the line." When the generator is "off the line," because the switch is at OFF or because of a momentary generator malfunction, the switch must be moved to RESET momentarily to bring the generator "on the line" and then released to ON to maintain the generator "on the line."

NOTE

Neither generator will operate unless the APU for the respective generator is also operating and driving the generator.

To bring the No. 1 generator "on the line" initially, the switch must be moved from OFF to RESET momentarily and then released to ON. It is not necessary to move the switch to OFF when the No. 1 APU is shut down, because the No. 1 generator underfrequency protective relay will have tripped the generator off.

NOTE

If either generator is operating and "on the line," most external power is automatically disconnected. Power to the battery bus, certain heaters, ready-to-launch light, liquid oxygen level probe, and stabilization of stable platform remains on.

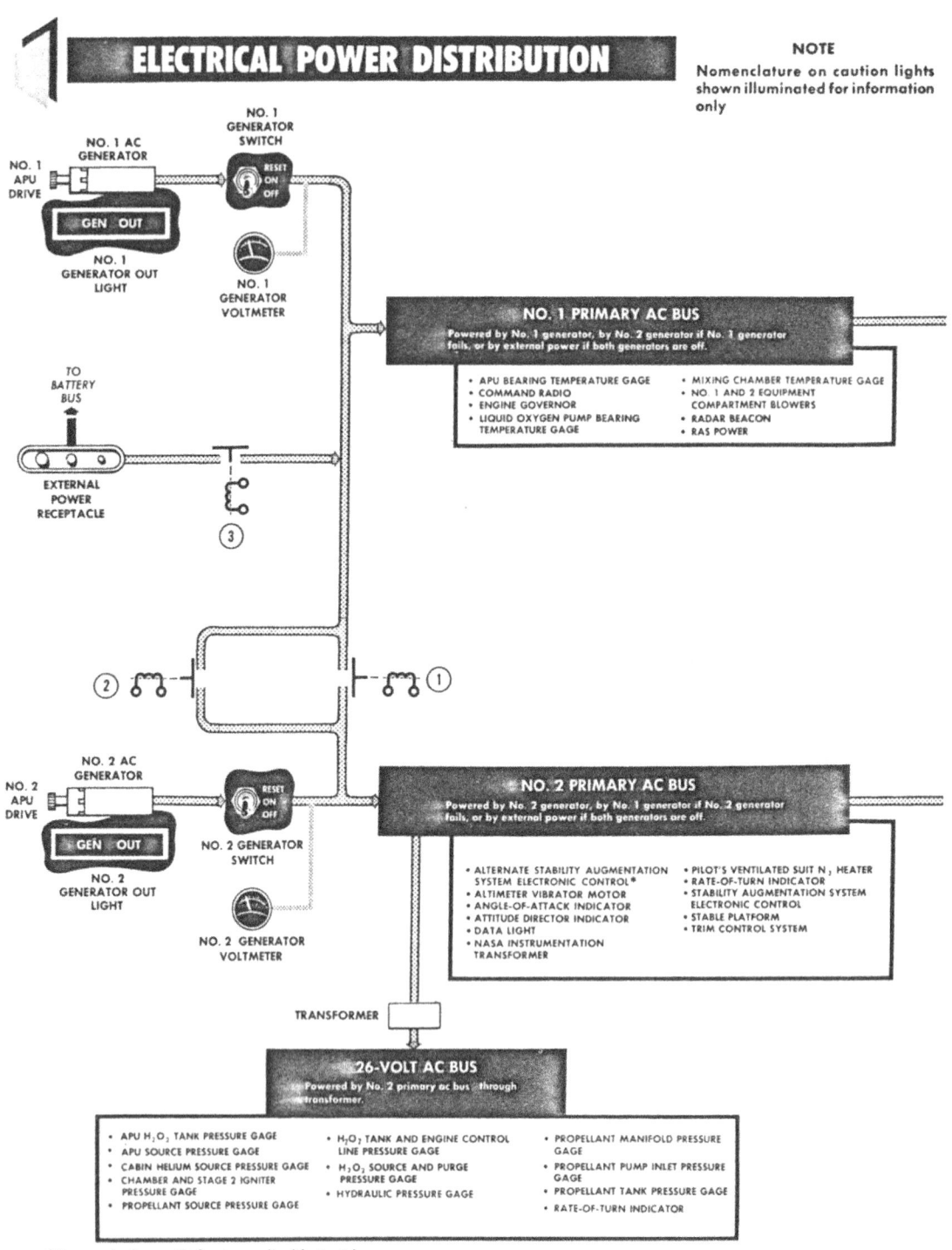

Figure 1-9

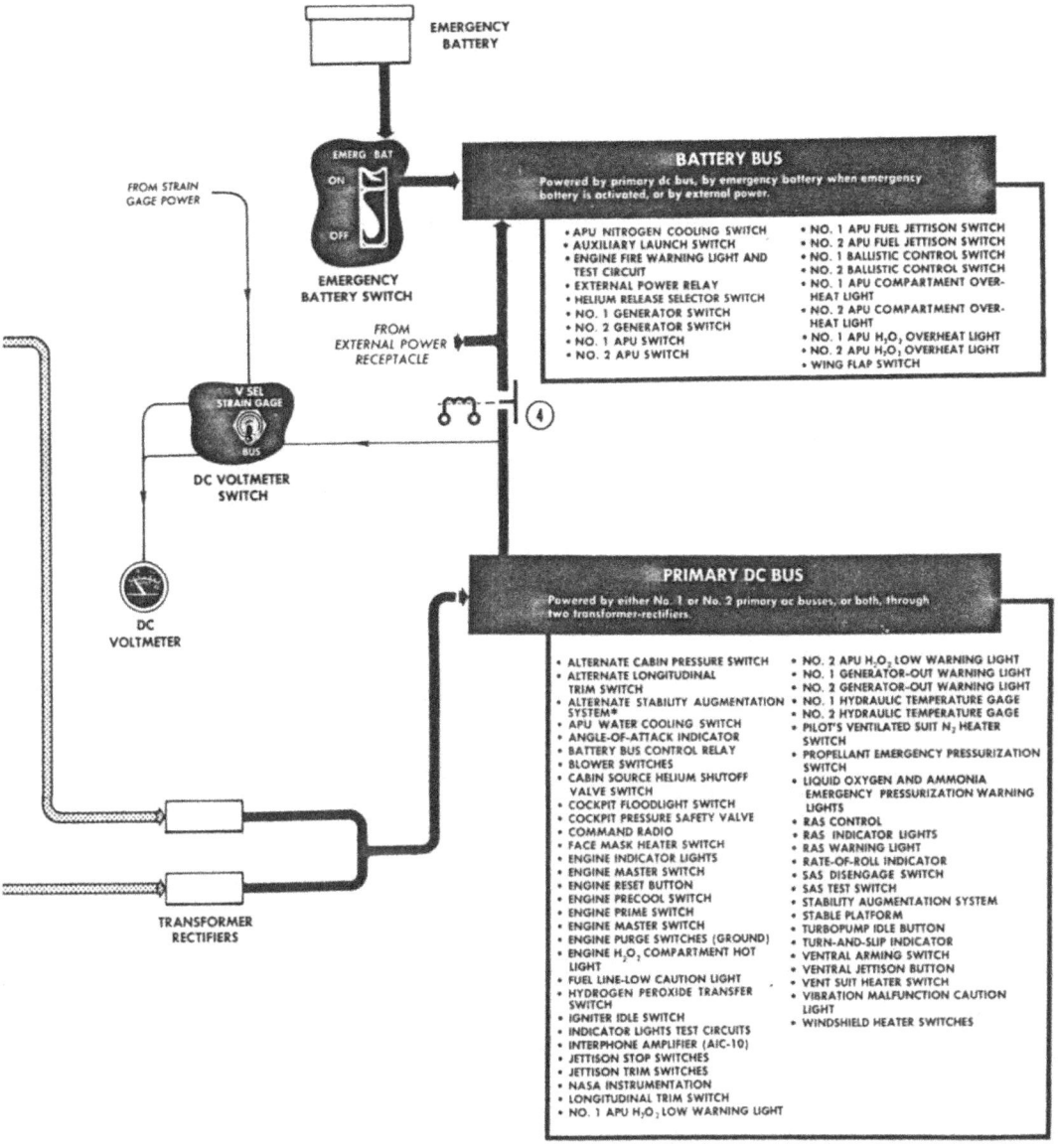

Section I   T.O. 1X-15-1

No. 2 Generator Switch.

Switch positions and operation of this switch (17, figure 1-2), on the instrument panel, are identical to those for the No. 1 generator control switch, except that, obviously, the No. 2 switch controls the No. 2 generator.

Emergency Battery Switch.

This guarded, two-position switch (16, figure 1-2), on the instrument panel, controls connection of the emergency battery to the battery bus. Normally, the switch is in the guarded OFF position. Raising the guard and moving the switch to ON connects the emergency battery directly to the battery bus.

DC Voltmeter Switch.

This three-position switch (8, figure 1-5), on the center pedestal, allows strain gage or primary dc bus voltage to be monitored. When the switch is at OFF, the dc voltmeter is disconnected. Moving the switch to STRAIN GAGE connects the dc voltmeter to the strain gage power. When the switch is at BUS, the voltmeter is connected to the primary dc bus.

ELECTRICAL POWER SUPPLY SYSTEM INDICATORS.

No. 1 Generator Voltmeter.

An ac voltmeter (11, figure 1-2), on the instrument panel, shows the No. 1 generator voltage. The instrument has a range of 0 to 250 volts, graduated in increments of 20 volts. The voltmeter reads line-to-line, rather than line-to-neutral or line-to-ground. Thus, the reading under normal conditions should be 200 volts and would check availability of two phases rather than only one phase. (When both primary ac busses are energized by one generator or by external power, both generator voltmeters should indicate the same voltage).

NOTE

If either generator fails, both the No. 1 and No. 2 generator voltmeters will show the output of the remaining generator.

No. 2 Generator Voltmeter.

An ac voltmeter (19, figure 1-2), on the instrument panel, shows the No. 2 generator voltage. The instrument has a range of 0 to 250 volts, graduated in increments of 20 volts. The voltmeter reads line-to-line, rather than line-to-neutral or line-to-ground. Thus, the reading under normal conditions should be 200 volts and would check availability of two phases rather than only one phase. (When both primary ac busses are energized by one generator or by external power, both generator voltmeters should indicate the same voltage.)

NOTE

If either generator fails, both the No. 1 and No. 2 generator voltmeters will show the output of the remaining generator.

No. 1 Generator-out Light.

Whenever the No. 1 generator drops "off the line," the amber No. 1 generator-out caution light (14, figure 1-2), on the instrument panel, comes on and reads "GEN OUT." The light is powered by the 28-volt primary dc bus.

No. 2 Generator-out Light.

Whenever the No. 2 generator drops "off the line," the amber No. 2 generator-out caution light (18, figure 1-2), on the instrument panel, comes on and reads "GEN OUT." The light is powered by the 28-volt primary dc bus.

DC Voltmeter.

The dc voltmeter (7, figure 1-5), on the center pedestal, has a range of 0 to 30 volts, graduated in increments of one volt. This voltmeter will indicate dc battery bus and strain gage power voltage. The voltage reading selection is through the dc voltmeter switch.

HYDRAULIC POWER SUPPLY SYSTEMS.

The airplane has a No. 1 and a No. 2 hydraulic system. (See figure 1-10.) Airplanes AF56-6670 and -6671 also have an SAS (stability augmentation system) emergency hydraulic system. The No. 1 and No. 2 systems are independent, but operate simultaneously to supply hydraulic pressure to all hydraulically operated systems of the airplane. Fluid is supplied to each hydraulic system by a reservoir, and pressure for each system is maintained by a variable high-displacement pump and a constant low-displacement pump. Each pump is driven through a gear train from the APU. (Refer to "Auxiliary Power Units" in this section.) The hydraulic system No. 1 pump is driven by APU No. 1; the system No. 2 pump is driven by APU No. 2. The hydraulic systems supply power for operation of the aerodynamic flight control system, speed brakes, and wing flaps. Dual, tandem hydraulic actuators are used, so that failure or shutdown of one hydraulic system will still permit the other hydraulic system to operate the various units. Each hydraulic actuator is capable of holding half of the maximum design hinge moment during single-system operation, which is adequate for control and landing of the airplane. The SAS pitch-roll servo cylinders are powered by the No. 2 hydraulic system. The SAS emergency hydraulic system will provide pressure to the pitch-roll servo cylinders in case of No. 2 system failure or shutdown of APU No. 2. The stability augmentation system yaw servo cylinder is powered only by hydraulic system No. 1. The No. 1 or No. 2 hydraulic system is automatically in operation whenever its respective APU is operating.

SAS EMERGENCY HYDRAULIC SYSTEM.

The SAS emergency hydraulic system (figure 1-10), on Airplanes AF56-6670 and -6671, provides hydraulic power for the SAS pitch-roll servo cylinders in case of failure of the No. 2 hydraulic system or its related APU. The 3000 psi system consists of a hydraulic motor, powered by the No. 1 hydraulic system, which drives a variable-displacement, constant-pressure hydraulic pump. A pressure-operated selector valve directs No. 1 hydraulic system pressure to the hydraulic motor in case of No. 2 system failure. The emergency system has its own reservoir.

HYDRAULIC PRESSURE GAGE.

A dual-movement, synchro-type gage (34, figure 1-2), on the instrument panel, indicates pressure in the No. 1 and No. 2 hydraulic systems. There are two pointers, numbered 1 and 2, for indicating respective system pressure. The gage has a range of 0 to 4000 psi in increments of 100 psi. The pressure indicating system receives power from the 26-volt ac bus.

HYDRAULIC TEMPERATURE GAGES.

Both the No. 1 and No. 2 hydraulic systems have a gage (24 and 45, figure 1-2) which indicates hydraulic fluid temperature. The gages are calibrated in degrees centigrade through the range of -100° C to 300° C in increments of 20° C. The temperature indicating systems receive power from the primary dc bus.

FLIGHT CONTROL SYSTEMS.

The airplane has two control systems. The aerodynamic flight control system (figure 1-10) consists of a mechanical system including hydraulically actuated control surfaces for use at altitudes where these surfaces are effective for maneuvering the airplane. The ballistic control system (figure 1-8) is used to control the airplane attitude at altitudes where the aerodynamic surfaces are relatively ineffective. The ballistic control system uses a monopropellant which is released through rockets at high velocity to rotate the airplane about its pitch, roll, and yaw axes as required for re-entry, and to correct oscillation.

AERODYNAMIC FLIGHT CONTROL SYSTEM.

The aerodynamic flight control system incorporates hydraulically actuated yaw and pitch-roll control surfaces. The irreversible characteristics of the hydraulic system hold the control surfaces against any forces that do not originate from pilot control movement and prevent these forces from being transmitted back to the pilot controls. Thus, aerodynamic loads of any kind cannot reach the pilot through the controls. An artificial-feel system is built into the control system to simulate feel at the pilot controls. In-flight trimming in pitch is accomplished by changing the neutral (no-load) position of the artificial-feel system and repositioning the control sticks. Yaw control is provided by movable upper and ventral vertical stabilizers. The left and right horizontal stabilizers provide pitch and roll control, simultaneous operation for pitch control, and differential operation for roll control. On Airplanes AF56-6670 and -6671, an assist to the aerodynamic damping in pitch, roll, and yaw is provided by a stability augmentation system (SAS).

Flight Control Hydraulic Systems.

When both hydraulic systems fail, the aerodynamic control surfaces will remain in the position at which failure occurred. However, the surfaces may be moved in the direction in which they are driven by aerodynamic loads by repositioning of the pilot controls in the direction to streamline the surface. The pedals, center stick, and console stick are mechanically linked to the control valves on their respective actuators. Movement of a pilot control results in corresponding movement of its actuator control valve. As the actuator moves, the control valve is repositioned to a neutral position so that flow to the actuator is shut off. The pressure remaining in the actuator serves to hold the control surface in the desired position. Control cable rig tension is maintained throughout a wide temperature and deflection range by thermal expansion and contraction tension regulators.

Artificial-feel and Trim Systems.

The artificial-feel system gives a sense of control feel to the pilot under all flight conditions where the aerodynamic controls are used. Aerodynamic stick and rudder pedal forces are simulated by spring-loaded bungees in the control system. The bungees apply loads to the pilot controls in proportion to stick or pedal movement, but the resultant feel has no relation to actual air loads. A nonlinear stick-to-stabilizer displacement ratio is incorporated in the pitch control linkage to minimize sensitivity. Pitch trim is obtained by shifting the neutral "no-load" position of the feel bungee to a stick position corresponding to the desired horizontal stabilizer position. Roll and yaw trim is adjustable only on the ground to compensate for airplane asymmetrical conditions.

Horizontal Stabilizer (Roll-Pitch Control).

The horizontal stabilizer consists of two all-movable, one-piece surfaces which can be moved simultaneously, differentially, or in compound. Aerodynamic control in pitch is obtained by simultaneous displacement of the left and right stabilizer surfaces. Roll control is obtained by differential displacement of the stabilizer surfaces. Combined pitch-roll control is obtained by compound movement of the stabilizer surfaces. A series of mixer bell cranks sum pilot control and SAS inputs to the two stabilizer actuator valves to obtain the desired pitch or roll-pitch control surface displacement.

Vertical Stabilizers (Yaw Control).

Aerodynamic control in yaw mode is obtained through displacement of the upper and ventral vertical stabilizers, which are actuated simultaneously through the coupled linkage between the upper and ventral vertical stabilizer actuator valves. Pilot pedal displacement and SAS yaw inputs are transmitted by mechanical linkage and cables to the synchronized upper and ventral stabilizer control valves. Since the ventral extends below the main landing skids, it must be jettisoned before landing. Four explosive bolts and a piston containing an explosive charge are electrically fired when the ventral jettison button is depressed. If this is not done, or if the ventral fails to jettison when the button is depressed, the ventral will jettison automatically when the landing gear is lowered. For either method of jettisoning, however, the ventral arming switch must be at ARM, to arm the jettison circuits.

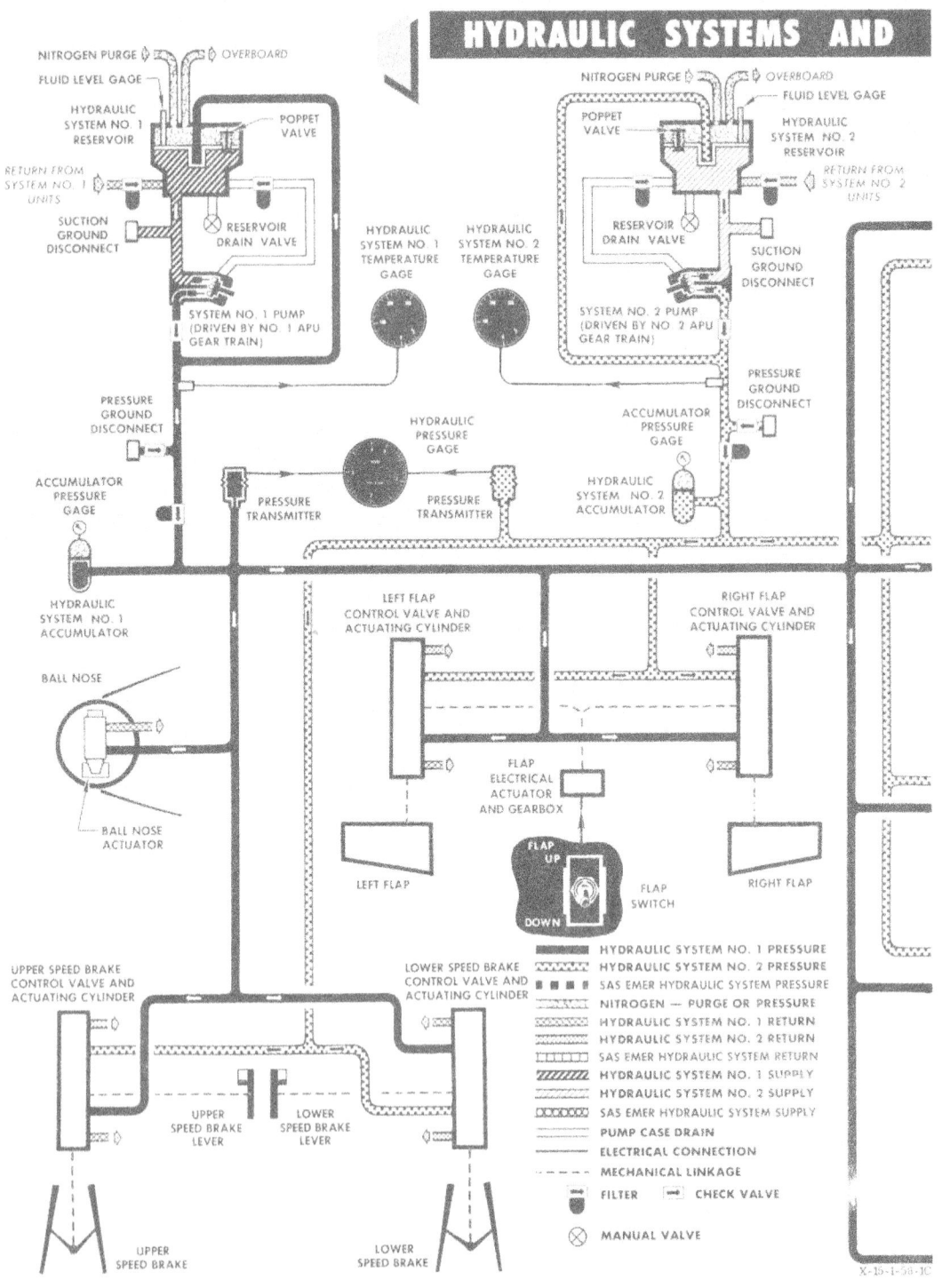

Figure 1-10. (Sheet 1 of 3)

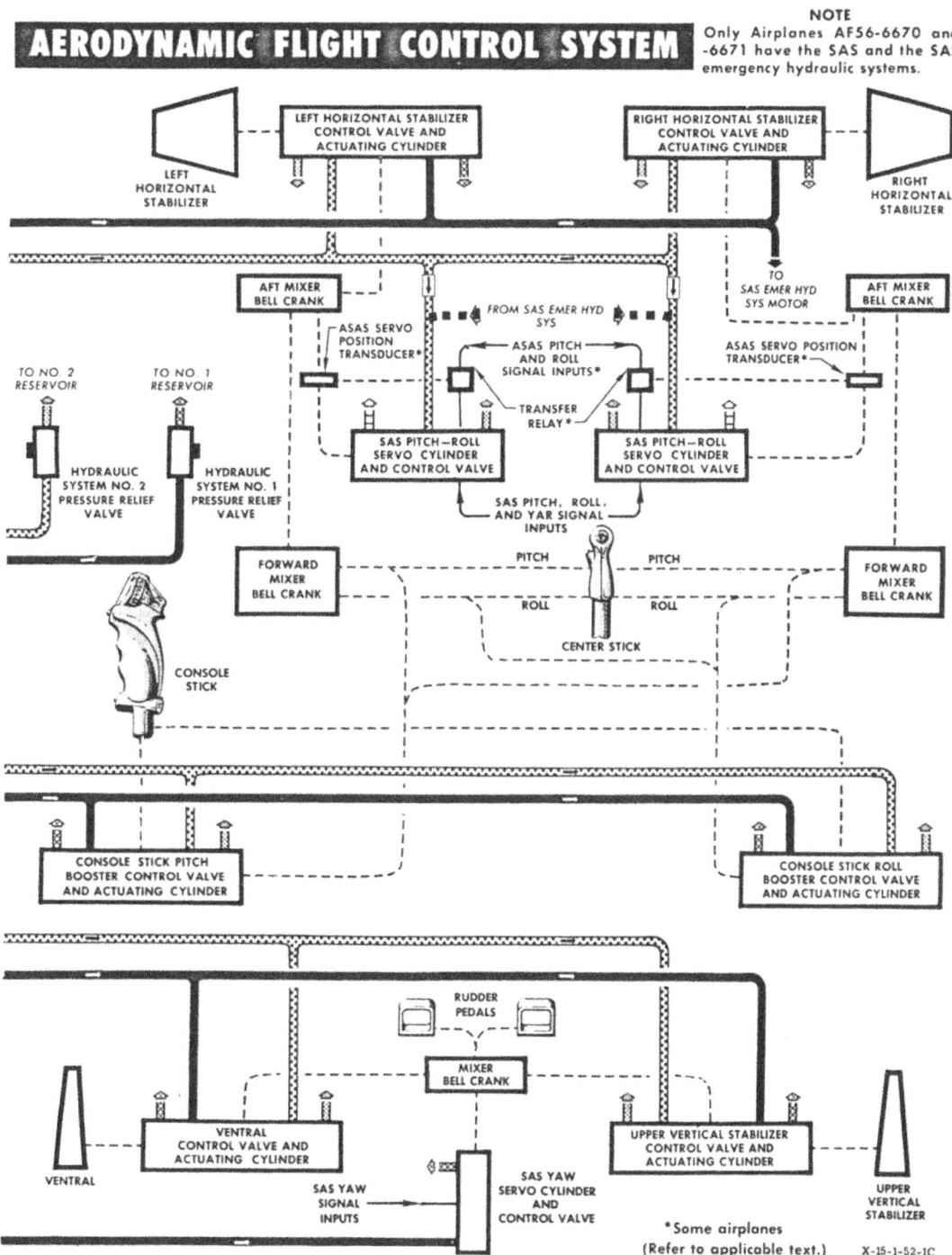

Figure 1-10. (Sheet 2 of 3)

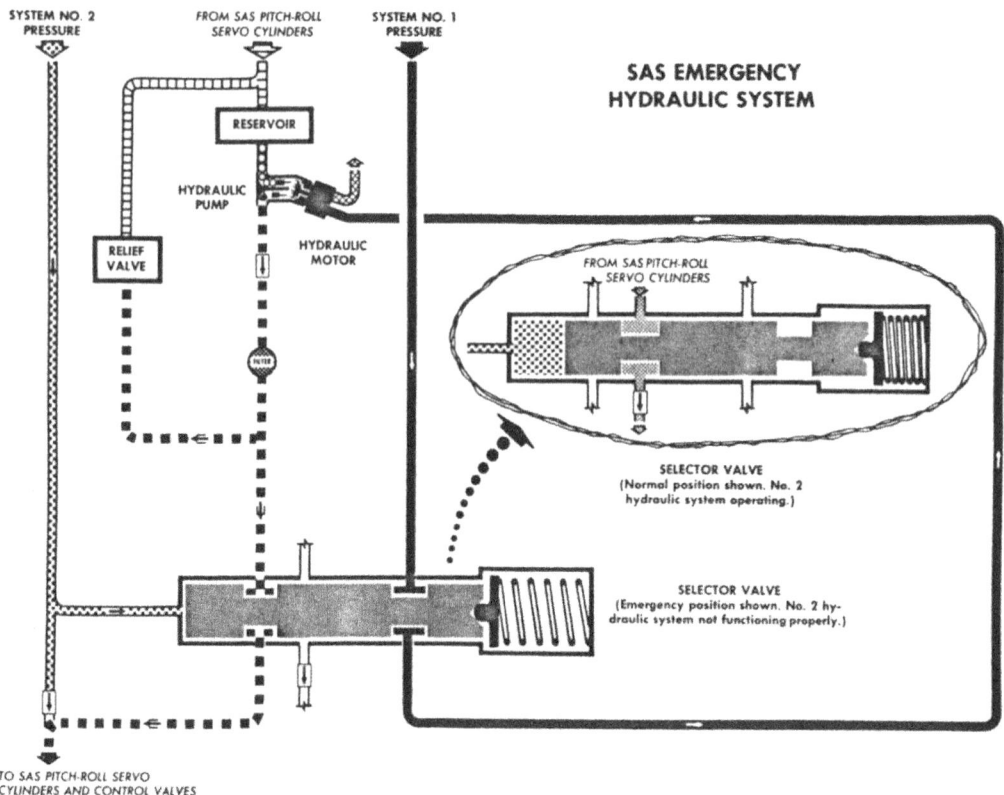

Figure 1-10. (Sheet 3 of 3)

Aerodynamic Flight Controls and Indicator.

Center Stick. The center stick (figure 1-11) is designed for use during normal periods of longitudinal and vertical acceleration. Pilot pitch and roll inputs to the center stick are summed by the mixer bell cranks and applied to the horizontal stabilizer actuator valves. A microphone button and alternate trim switch are on the stick grip. The button is in parallel with the microphone button on the console stick.

Console Stick. The console stick (figure 1-11), on the right console, enables the pilot to control the airplane throughout the periods of high longitudinal and vertical accelerations. This stick has full range of surface control in pitch and roll and is coupled to the center stick linkage through separate pitch and roll hydraulic boost actuators to reduce console stick pilot control forces and to synchronize displacement of the center and console sticks. The console stick has a pitch trim knob and a microphone button. The trim knob is graduated in degrees ranging from ±3 to -25 degree stabilizer leading edge travel. When the trim control switch is in NORMAL, moving the trim knob causes corresponding movement of the pitch trim actuator. A microphone button on the console stick is in parallel with the microphone button on the center stick.

Rudder Pedals. Conventional rudder pedals, which are adjustable, are mechanically linked to the yaw system mixer bell crank. Pedal movement and SAS inputs are summed by a mixer bell crank, which in turn transmits the summed signal mechanically to the stabilizer actuator control valves.

Horizontal Stabilizer Position Indicator. A horizontal stabilizer position indicator (figure 1-11), on the right vertical side panel, provides a quick reference to the position of the horizontal stabilizer before re-entry. The forward end of the inboard face of the indicator is labeled "DIVE"; the aft end, "CLIMB." On the top of the indicator is a scale calibrated in increments of 5 degrees, from 0 to 15 degrees stabilizer leading-edge up, and from 0 to 35 degrees stabilizer leading-edge down. A red index marker is attached to the console stick shaft so that as the stick is moved, the marker will point to the corresponding horizontal stabilizer position on the scale.

FLIGHT CONTROLS AND INDICATORS

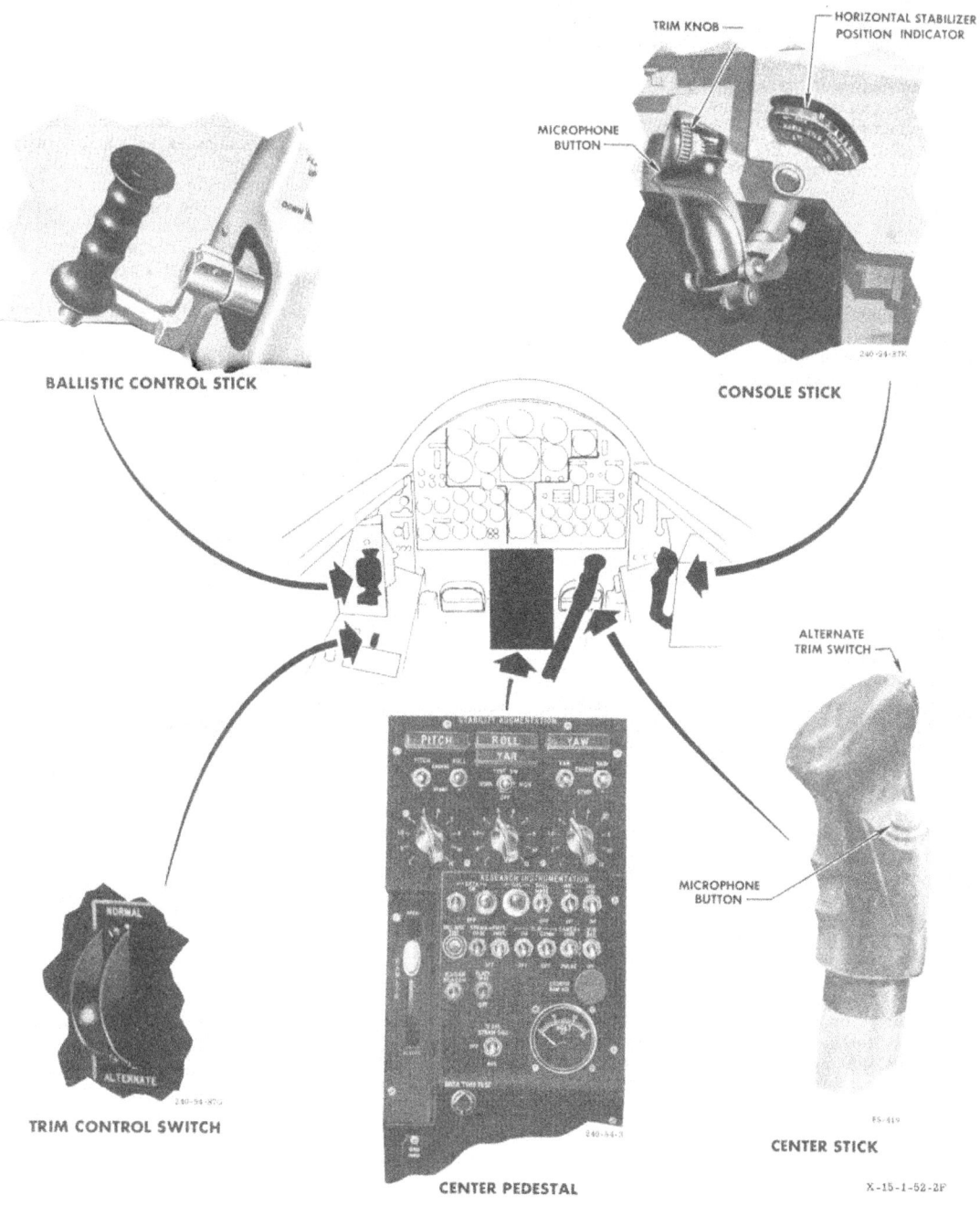

Figure 1-11

Trim Control Switch. This two position "LONG TRIM" switch (1, figure 1-3), on the left console, controls selection of normal or alternate trim. With the switch at the maintained NORM position, the center stick trim control knob is inoperative, and primary ac and dc power is applied to the electronic trim amplifier and relay unit to activate the console stick pitch trim knob. Movement of the pitch trim knob on the console stick then causes primary dc power to be applied to the pitch trim actuator. With the trim control switch at the maintained ALT position, output of the electronic trim amplifier is cut off, the pitch trim knob is inoperative, and pitch trim is accomplished through the alternate trim switch on the center control stick.

Alternate Trim Switch. This three-position switch on the center stick grip (figure 1-11) controls primary dc bus power to the pitch trim actuator. The switch is spring-loaded to the center position, and all positions are unmarked. With the switch in the center position and the trim control switch at NORM, primary dc bus power is applied to the electronic trim amplifier and relays, and pitch trim is accomplished by movement of the pitch trim knob. Moving the alternate trim switch forward (NOSE DOWN) or aft (NOSE UP) bypasses the electronic trim amplifier and applied primary dc bus power directly to the pitch trim actuator, causing the actuator to move for nose down or nose up trim, as selected.

CAUTION

If the trim control switch is moved from ALT to NORM, the pitch trim actuator will run to the preset position of the pitch trim knob.

Ventral Arming Switch. This guarded, two-position switch (28, figure 1-2) is on the instrument panel right-hand wing and is labeled "VENTRAL." With the switch at DE-ARM, the circuit which controls primary dc bus power to the explosive bolts and explosive charge in the ventral jettison piston is interrupted and the ventral cannot be jettisoned either selectively through the ventral jettison button or automatically when the landing gear is lowered. With the switch at ARM, the ventral jettison circuit is armed.

Ventral Jettison Button. This jettison button (68, figure 1-2), labeled "AUX VENTRAL JETT," is recessed within a circular guard on the instrument panel left wing. The button is powered by the primary dc bus and is the primary means of jettisoning the ventral. When the button is depressed, with the ventral arming switch at ARM, four explosive-type bolts and a piston containing an explosive cartridge are fired. At the instant the bolts separate, the piston is driven downward to forcibly jettison the ventral.

Hydraulic Pressure Gage. Refer to "Hydraulic Power Supply Systems" in this section.

STABILITY AUGMENTATION SYSTEM (SAS).

The stability augmentation system (SAS), installed on Airplane AF56-6670 and -6671 provides damping inputs to the aerodynamic flight control system about the pitch, roll, and yaw axes. (See figure 1-10.) Major components of the system are a three-axis gyro, an electronic case assembly, two pitch-roll servo cylinders, one yaw servo cylinder, and a gain selector switch assembly. In-flight testing of the SAS channels can be done by use of an in-flight test unit controlled from the cockpit. The SAS has fundamentally three semi-independent channels, each comprising a working circuit, a monitor circuit, and a malfunction detector. Each working circuit received its commands from the gyro assembly in its particular axis and in turn commands associated servo cylinder displacement. The monitor circuits receive commands from the gyro assembly identical to those of the control circuits. The malfunction detectors compare the commands passing through the monitor circuits with the associated servo cylinder displacement. If a predetermined error between the working and monitor circuits is exceeded, the malfunction detector locks out the particular channel in which the error occurred. The working circuit command to the associated servo cylinders is an electrical signal which drives an electro-hydraulic transfer valve on the servo cylinder. The transfer valve controls hydraulic pressure on each side of the servo cylinder piston. The pitch-roll servo cylinders, powered by the No. 2 hydraulic system, are mechanically linked to the horizontal stabilizer control linkage by mixer bell cranks and move surfaces by simultaneous movement of the bell cranks. Roll is achieved by differential movement of the bell cranks. In addition, loss of No. 2 hydraulic system pressure will automatically engage the SAS emergency hydraulic system. Should an abrupt loss of No. 2 system pressure occur, the pitch and roll damping channels may trip. If this occurs, they must be reset. The yaw servo cylinder, powered by the No. 1 hydraulic system, is mechanically linked to the vertical stabilizer control linkage through a mixer bell crank and moves surfaces by movement of the bell crank. An interaction of the yaw and roll damping working circuits is provided whereby signals from the yaw axis of the gyro are fed into the roll circuit to augment roll damping. This is referred to as the "yar" function. The SAS is powered from the No. 2 primary ac bus and the primary dc bus and is in ready status continuously as long as these busses are powered.

CAUTION

- The 28-volt dc SAS circuit breaker must be closed (pushed in) before the airplane is fueled, and must be kept closed until the end of the flight, so that the SAS gyro heaters will be energized to prevent freezing damage to the gyro

- Airplane AF56-6672 has an MH-96 adaptive flight control system installed for test purposes, in place of the SAS and reaction augmentation system. For information on the MH-96 system, refer to MH Aero Report No. 2373-TM1, Volumes I through VIII.

Stability Augmentation System Controls and Indicators.

All stability augmentation system controls and indicators are on the SAS control panel on the center pedestal.

Pitch Function Switch. This two-position switch, (17, figure 1-5), labeled "PITCH," controls the SAS pitch channel circuitry. Both switch positions are maintained. With the switch at STDBY, the pitch channel is functioning, but the input signals to the servo cylinder control valves and hydraulic pressure to the servo cylinders are shut off. With the switch at ENGAGE, pitch damping signals are applied to the pitch-roll servo cylinder control valves, which in turn permits hydraulic power to be applied to the servo cylinders. The switch is powered by the primary dc bus.

NOTE

If both the pitch and roll function switches are at STDBY, the pitch-roll servo cylinders are centered and locked.

Roll Function Switch. This two-position switch, (16, figure 1-5), labeled "ROLL," is powered by the primary dc bus. It controls the SAS roll channel circuitry. With the switch at STDBY, the roll channel is operating, but the input signals to the servo cylinder control valves and hydraulic pressure to the servo cylinders are shut off. With the switch at ENGAGE, roll damping signals are applied to the pitch-roll servo cylinder control valves, which in turn permits hydraulic power to be applied to the servo cylinders. Both switch positions are maintained.

NOTE

If both the pitch and roll function switches are at STDBY, the pitch-roll servo cylinders are centered and locked.

Yaw Function Switch. The yaw function switch (3, figure 1-5), labeled "YAW," has two maintained positions. The switch is powered by the primary dc bus. With the switch at STDBY, the SAS yaw channel is operating, but the input signals to the servo cylinder control valve and hydraulic pressure to the servo cylinder are shut off and the yaw servo cylinder is centered and locked. With the switch at ENGAGE, yaw damping signals are applied to the yaw servo cylinder control valve, which in turn permits hydraulic power to be applied to the servo cylinder.

NOTE

Shutoff or failure of hydraulic system No. 1 automatically causes the yaw servo cylinder to recenter and lock.

"Yar" Function Switch. This two-position switch, (2, figure 1-5), labeled "YAR," is powered by the primary dc bus. The switch controls the yaw signal input to the SAS roll control circuit. With the switch at STDBY, the "yar" signal circuit is inoperative. With the switch at ENGAGE while the roll function switch is at ENGAGE, yaw signals are applied to the SAS roll control circuit. Both switch positions are maintained.

NOTE

If the roll function switch is at STDBY or the roll control circuit does not function properly, yaw input to the SAS roll control circuit will neutralize roll and result in a "no roll" output.

Gain Selector Knobs. Three gain selector knobs (4, 13, and 14, figure 1-5) control, through selection of fixed resistors, the ratio of pitch, roll, and yaw damping signal to servo cylinder displacement. There are 10 switch positions; the third position is designated as LO, and the ninth is designated as HI. Pitch gain is controlled by the left knob, roll gain by the center knob, and yaw gain by the right knob. For a given gyro signal, servo cylinder displacement is lowest with a knob at position 1, and the highest at position 10.

NOTE

Gain selector knob positions to be used in flight depend on damping requirements the pilot considers necessary. However, the initial setting depends on the particular flight conditions.

SAS Caution Lights. Four placard-type amber caution lights (1, figure 1-5) indicate operating status of the SAS control circuits. The lights are powered by the primary dc bus and can be tested through the indicator, caution, and warning light test circuit. There is one light for each of the pitch, roll, yaw, and "yar" channels. When the lights are on, they read "PITCH," "ROLL," "YAW," and "YAR," respectively. The pitch, roll, and yaw lights are on when the pitch, roll, and yaw function switches are at STDBY. When any one of these switches is at ENGAGE and its control circuit is operating normally, the associated caution light is out. When the circuit error exceeds the predetermined limit, the associated caution light blinks at approximately a 4-cycle-per-second rate, and the affected channel is automatically disabled. If the error returns to within limits, the function switch must be moved to STDBY to reset the channel and then back to ENGAGE in order to restore the channel to operation. If the error is still out of limits, the function switch may be returned to STDBY and the caution light will be constantly on. The "YAR" caution light shows only the "yar" function switch position. When the switch is at STDBY, the light is on. When the switch is at ENGAGE and the "yar" signal is available to the roll channels, the light is out. Any error in the "yar" circuit exceeding the established limits is monitored by the roll channel malfunction detector and causes the roll caution light to come on.

SAS Test Switch. The three-position SAS test switch (15, figure 1-5) is on the center pedestal. The switch is powered by the primary dc bus. It has momentary WORK and MON positions and a spring-loaded OFF position. While the pitch, roll, and yaw damping channels are engaged, singly or in combination, their working and monitor circuits can be tested during captive or free flight. Placing the switch at either WORK or

Section I

T.O. 1X-15-1

MON opens the associated rate gyro ground circuits (working or monitor) and inserts a calibrated test voltage in series with the pick-offs. This voltage is added to the normal output of the gyro and, if the gain selector knobs are set to prespecified positions, the test voltage unbalances the SAS channels beyond the expected trip level of the SAS malfunction detectors. If the SAS is functioning properly, the SAS caution lights for the channels being tested will blink, signifying malfunction circuit operation. After each set of circuits (working or monitoring) is tested, the SAS channels must be re-engaged.

NOTE

The "yar" function switch must be at STDBY during the tests.

SAS Disengage Switch.

This two-position switch (15, figure 1-3), labeled "LAND SAS DISENGAGE," is used to arm the SAS main gear cutout relay to shut off the SAS upon main gear touchdown. When the switch is moved forward to the LAND SAS DISENGAGE position, primary dc bus power is applied to the cutout relay to arm it. Moving the switch aft to the unmarked off position turns off power to the relay.

ALTERNATE STABILITY AUGMENTATION SYSTEM (ASAS).*

The alternate stability augmentation system (ASAS) (figure 1-10) provides damping inputs to the aerodynamic flight control system about the pitch and roll axes in case of failure of the roll axis of the SAS. Major components of the system are the pitch and roll gyros, electronic case assembly, left and right transfer relays, and left and right servo position transducers, in addition to the servo cylinders and bell cranks which are common to both the SAS and the ASAS. When the ASAS is armed, failure of the roll channel in the SAS will automatically energize the ASAS. Change-over to the ASAS does not affect the yaw channel of the SAS, which continues to function in the normal manner. The ASAS is completely separate from the SAS and comprises a working circuit only, bypassing all SAS controls and indicators which are then inoperative (except for those in the yaw channel). To completely bypass the SAS electrical circuitry, a servo position transducer, separate from those on the SAS servo cylinders, is on the left and right SAS bell cranks to transmit servo cylinder position feedback to the ASAS gyros and electronic case assembly. The ASAS is powered from the No. 2 primary ac bus and the primary dc bus.

NOTE

The ASAS will be used as directed in the mission profile.

Alternate Stability Augmentation System Controls and Indicators.

Alternate SAS Switch. This two-position switch (10, figure 1-3), labeled "ALT SAS," permits automatic change-over to the ASAS at pilot option. When the switch is at OFF, the system has power applied and is in a stand-by condition. When the switch is moved to the ARM position, a malfunction in the roll channel of the SAS, or placing of the SAS roll function switch at STDBY, will cause the transfer relay to be energized, thereby connecting the ASAS input signals to the pitch-roll servo cylinders and disconnecting the SAS input signals.

Alternate SAS On Light.

A placard-type green indicator light (5A, figure 1-2), labeled "ALT SAS ON," informs the pilot when the ASAS is engaged and operating.

BALLISTIC CONTROL SYSTEM.

The ballistic control system is used to control airplane attitude at flight altitudes where aerodynamic flight controls are ineffective. Ballistic control is provided by two independent systems which normally are operated simultaneously. Each system uses a monopropellant which is converted by catalytic action to superheated steam and oxygen and is released through small rockets in the nose section and wings. Figure 1-8 illustrates schematically the operation of one system of the ballistic controls. Operation of the other system is identical. The reaction of the escaping gas causes the airplane to move about the selected axis or combination of axes. The monopropellant, hydrogen peroxide, is supplied from the APU and ballistic control propellant feed system. (Refer to "APU and Ballistic Control Propellant Feed Systems" in this section.) The hydrogen peroxide tank which supplies the No. 1 APU also supplies the No. 1 system of the ballistic controls. The tank which supplies the No. 2 APU also supplies the No. 2 system of the ballistic controls. Movement of the ballistic control stick in the cockpit opens a metering valve, allowing the monopropellant to enter the selected rockets. In the rockets, the hydrogen peroxide enters catalyst chambers, where it is decomposed into a high-pressure gas mixture of superheated steam and oxygen. The gases then exhaust through the nozzles. The reaction of the escaping gases causes the airplane to move about the selected axis in a direction opposite to that of the escaping gases. There are six rockets in each system. One system includes four rockets in the nose and the two left wing rockets. The other system includes the remaining four rockets in the nose and the two right wing rockets. A dual metering valve controls flow of the monopropellant to the eight nose rockets for pitch and yaw, and a dual metering valve controls flow of the monopropellant to the four wing rockets for roll. Flow of the monopropellant to the metering valves is controlled by two switches in the cockpit, one for each system. With both systems operating, a nose-down selection from the cockpit causes operation of the two rockets in the

1-34

Changed 26 April 1963

top of the nose section. A nose-right selection from the cockpit causes operation of the two rockets in the left side of the nose. A right-roll selection from the cockpit causes operation of the rocket whose nozzle is in the bottom of the left wing and the rocket whose nozzle is in the top of the right wing. Stick force gradients are maintained for all three axes of operation by spring bungees. For the pitch and yaw axes, an increase in force versus deflection rate of the ballistic control stick marks half of the maximum control travel and half of the maximum force. Thus, the pilot feels the mid-point of maximum opening of the metering valve. The acceleration and velocity of airplane movement about an axis vary with the amount and duration of ballistic control stick application. The velocity tends to sustain itself after the stick is returned to the neutral position. A subsequent stick movement opposite to the initial one is required to cancel the original attitude change. The No. 1 ballistic control system rockets also are used by the reaction augmentation system, which is installed on Airplanes AF56-6670 and -6671. Refer to "Reaction Augmentation System (RAS)" in this section. If either ballistic control system fails, the other system provides adequate power to control the airplane. A transitional altitude band exists wherein it will be necessary to use the ballistic control system and the aerodynamic flight controls simultaneously for maneuvering and controlling airplane attitude. The size of this transitional band is somewhat affected by airplane speed and the amount of maneuvering or attitude change required.

Ballistic Control System Controls.

Ballistic Control Switches. The No. 1 and No. 2 ballistic control switches (36, figure 1-2), on the instrument panel, are powered by the primary dc bus. Moving either switch to ON simultaneously opens the helium shutoff valve and the propellant feed system jettison and ballistic control valve for the respective system and allows hydrogen peroxide to flow to the metering valves. With either switch at OFF, the propellant feed system jettison and ballistic control valve for the respective system is turned off. However, the helium shutoff valve will not close as long as either APU switch is at ON.

NOTE

When either APU switch is at JETT, the feed system jettison and ballistic control valve for the related system of the ballistic controls closes and shuts down that system.

Ballistic Control Stick. The ballistic control stick (figure 1-11), above the left console, is mechanically connected to the system metering valves. When the ballistic control system is operating, moving the stick positions the metering valves to allow flow of hydrogen peroxide to the selected rockets in the nose and wings. Yaw control is obtained by direct left or right movement of the stick. For left yaw, the stick must be moved directly to the left. Roll control is obtained by rotation of the stick. For left roll, the stick must be rotated to the left (counterclockwise). Pitch control is obtained by direct up or down movement of the stick. An airplane nose-up pitch change is obtained by raising the stick.

Nose Ballistic Rocket Heater Switch. The nose ballistic rocket heater switch (27, figure 1-2), on the instrument panel right-hand wing, receives electrical power from the carrier airplane. Moving this switch

THIS PAGE INTENTIONALLY LEFT BLANK.

REACTION AUGMENTATION SYSTEM CONTROLS

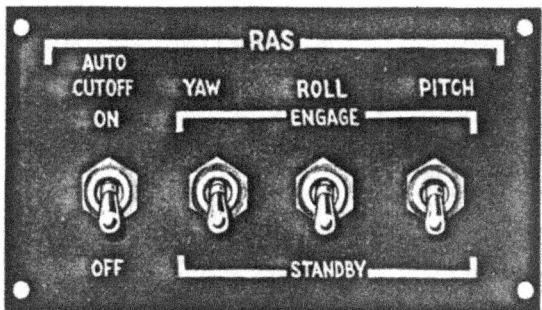
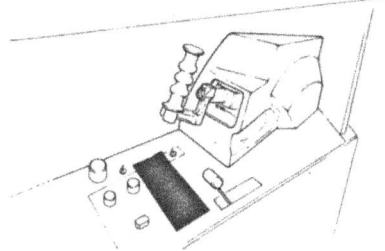

Figure 1-12

to ON during captive flight starts the electric heaters on the ballistic control system rockets in the nose of the X-15 to preheat the rockets. After launch, this switch is inoperative.

REACTION AUGMENTATION SYSTEM (RAS).

The reaction augmentation system (RAS) installed on Airplanes AF56-6670 and -6671, provides rate damping to aid airplane control and minimize pilot overcontrol when the ballistic control system is used. The RAS conserves APU and BCS H_2O_2 supply by minimizing pilot overcontrol tendencies at low acceleration levels. The RAS is incorporated in the No. 1 ballistic control system and uses the rocket motors associated with that system. Pilot control of both system No. 1 and No. 2 rocket motors is still maintained. Whenever the ballistic control stick is in neutral, the RAS gyro accelerometer assembly senses angular rate about the three axes and at a preset angular rate, a switching system is energized to fire the proper No. 1 system rocket motor to reduce the sensed angular acceleration. The RAS is automatically turned off by means of an acceleration-sensing device upon a build-up of G, which corresponds to an increase in the effectiveness of the aerodynamic controls. The RAS can be overridden by the pilot in case of a failure. The RAS is in a stand-by condition whenever power is applied to the airplane and is engaged by switches in the cockpit. Any combination of the three axes can be selected individually as desired. An indicator light in the cockpit shows when the RAS is in stand-by. Electrical power for the RAS is obtained from the No. 1 primary ac bus and primary dc bus. The system is protected by circuit breakers in the equipment bay.

Reaction Augmentation System Controls and Indicator.

All reaction augmentation system controls are on the RAS control panel (figure 1-12) on the left console aft of the ballistic control stick.

Automatic Cutoff Switch. This two-position switch on the RAS control panel is labeled "AUTO CUTOFF." Moving the switch to ON engages the automatic cutoff. With the switch at ON, automatic cutoff of the RAS will occur when a preset level of normal acceleration is sensed. When automatic cutoff occurs, an indicator light comes on. To reset the automatic cutoff feature, the switch is moved to OFF, then returned to ON. The OFF position of the switch disengages only the automatic cutoff feature.

Pitch Switch. The two-position "PITCH" switch on the RAS control panel has ENGAGE and STANDBY positions. Moving the switch to ENGAGE closes the pitch circuit and permits operation of the proper pitch rocket motor by the RAS. With the switch at STANDBY, the pitch output circuit is opened.

Roll Switch. The two-position "ROLL" switch on the RAS control panel has ENGAGE and STANDBY positions. Moving the switch to ENGAGE closes the roll circuit and permits the operation of the proper roll rocket motor by the RAS. With the switch at STANDBY, the roll output circuit is opened.

Yaw Switch. The two-position "YAW" switch on the RAS control panel has ENGAGE and STANDBY positions. Moving the switch to ENGAGE closes the yaw circuit and permits operation of the proper yaw rocket motor by the RAS. With the switch at STANDBY, the yaw output circuit is opened.

RAS Indicator Lights.

Three amber indicator lights (9, figure 1-3), labeled "YAW," "ROLL," and "PITCH," are powered by the primary dc bus and come on whenever the corresponding yaw, roll, and pitch switch is in the ENGAGE position. The lights are tested through the indicator, caution, and warning light test circuit.

RAS-out Indicator Light. The RAS-out indicator light (12, figure 1-2), on the instrument panel, receives primary dc bus from the gyro accelerometer assembly and comes on to show "RAS OUT" whenever automatic cutoff occurs of if the control switches of all three axes are at STANDBY. Moving any one of the three switches to ENGAGE turns off the light if the automatic cutoff circuit has not been tripped. If the light is on because of automatic cutoff, the automatic cutoff circuitry can be reset by moving the "AUTO CUTOFF" switch to OFF, then to ON, to put out the light. The light can be tested by means of the indicator, caution, and warnin light switch.

MH-96 ADAPTIVE FLIGHT CONTROL SYSTEM.

The MH 96 adaptive flight control system is installed for test purposes on Airplane AF56-6672. For information on description and operation of this system, refer to MH Aero Report 2373-TM1, Volumes I through VIII.

WING FLAP SYSTEM.

Flap position is controlled by an electromechanical actuator, containing two electric motors that are coupled together by a set of differential gears. The output of this actuator drives a push-pull cable system which opens the valves in the dual, tandem hydraulic flap actuators and hydraulically positions the flaps to either full up or full down. Flap extension is possible even if one motor fails; however, the extension time is approximately twice the normal time. Normal extension from full up to full down position requires about 8 to 10 seconds. No provisions have been made for automatic pitch correction with flap extension, nor for intermediate positioning of the flaps. However, because of the incorporation of hydraulic relief valves to limit the maximum air load on the surfaces, the flaps may partially close at speeds above 250 knots. No flap position indicator is provided.

WING FLAP SWITCH.

A two-position switch (8, figure 1-3) is on the left vertical side panel. The switch, labeled "FLAP," controls flap operation, and is powered by the battery bus. It has two positions, UP and DOWN. No intermediate positioning is provided.

SPEED BRAKE SYSTEM.

The airplane has two speed brakes, one on the fixed portion of the upper vertical stabilizer and the other on the fixed lower vertical stabilizer. Each speed brake consists of two symmetrical panels, hinged at the forward end. Each speed brake is operated by a dual, tandem hydraulic actuator. One segment of an actuator is powered from hydraulic system No. 1; the other segment, by hydraulic system No. 2. Failure of one hydraulic system still permits operation of the speed brakes, however, at reduced rate under any particular air load. The actuator valves are controlled from the cockpit by a system of cables and mechanical linkage. Follow-up mechanisms permit positioning each speed brake to any position between fully closed and fully open.

NOTE

Each speed brake actuator incorporates a relief valve which prevents the speed brake from extending or allows the speed brake to retract under excessive air loads, to prevent structural damage.

WARNING

The speed brakes on this airplane were not designed for use as a low-speed drag device; opening them at subsonic speeds can be dangerous. Their design function is to provide necessary drag conditions for control of the airplane at supersonic speeds and relatively high altitudes.

SPEED BRAKE HANDLES.

The speed brake handles (13 and 14, figure 1-3) are on the left console. The inboard handle controls the lower speed brake; the outboard handle controls the upper speed brake. The handles normally are locked together by an interconnecting bolt at the forward cable sectors to ensure symmetrical operation of the speed brakes. A spring-loaded lock lever on the inboard handle is designed to unlock the handles for independent speed brake operation and lock them for symmetrical operation if the interconnecting bolt is not installed. Speed brake position is indicated by a scale for each handle on the speed brake handle quadrant. The scales are calibrated in increments of 5 degrees. When the handles are moved to a given setting, the speed brakes will open to the position selected.

LAUNCH SYSTEM.

The dropping of this airplane from the carrier airplane is normally performed by the pilot of the carrier airplane. However, if this cannot be done, the captive airplane is equipped with an auxiliary launch switch to perform this function from a separate power source. A switch in the captive airplane controls an indicator light in the carrier airplane to indicate that the pilot of the captive airplane is ready for launch.

READY-TO-LAUNCH SWITCH.

The ready-to-launch switch (57A, figure 1-2), on the instrument panel, is moved up (ON) to turn on the "READY TO LAUNCH" light in the carrier airplane to indicate when the X-15 is ready to be dropped. There is no indicator light in the X-15. Moving the "READY TO LAUNCH" switch down (OFF) turns off the indicator light. This circuit is powered by the carrier airplane's electrical system.

AUXILIARY LAUNCH SWITCH.

The guarded, battery-bus-powered auxiliary launch switch (66, figure 1-2), on the instrument panel left wing, uses primary dc bus power to operate the hydraulic launch system in the carrier airplane. Lifting the guard and moving the switch up (ON) supplies power directly to the solenoid valve in the normal launch hydraulic system.

LANDING GEAR SYSTEM.

The two main landing gears are of the skid type and lie adjacent to the lower aft fuselage and parallel to the airplane centerline when retracted. The skids (27, figure 1-1) are mounted on inflexible struts with an air-oil shock absorber attached to the upper end which permits some outward rotation when the weight of the airplane is on the landing gear. The nose gear (3, figure 1-1) is a conventional, nonsteerable, dual-wheel type and retracts forward, fairing into the fuselage nose section. Both the main and nose gear, when unlocked, extend by gravity and air loads. However, the nose gear lowering system includes an initiator to ensure positive nose gear lowering. They cannot be retracted by the pilot. No gear-down indication is provided. Gear retraction must be accomplished manually by ground personnel. When the main gear is released, a microswitch on the left main gear activates an explosive charge, causing the ventral to jettison, provided the ventral arming switch is at ARM and if the ventral had not been previously jettisoned by use of the ventral jettison button.

LANDING GEAR HANDLE.

The T-type landing gear handle (67, figure 1-2) is on the instrument panel left wing. The handle is mechanically linked to the main gear uplocks and the nose gear and nose gear door uplocks. When the handle is pulled straight aft approximately 11 inches, the uplocks are released, the spring-loaded scoop door in the nose gear door swings downward into the airstream, and the nose gear extension initiator fires. Gravity and air loads cause the main gear to extend and lock. Air loads on the nose gear scoop door force the nose gear down and locked. (The initiator actuates a piston which forces the nose gear door open under flight attitudes where air loads tend to hold the door closed.)

The landing gear handle should be manually stowed to prevent possible damage to the instrument panel.

INSTRUMENTS.

Most of the instruments are powered by the ac or dc electrical systems or a combination of both.

NOTE

For information regarding instruments that are an integral part of a particular system, refer to applicable paragraphs in this section and Section IV.

PITOT-STATIC SYSTEM.

Pitot pressure for the conventional airspeed indicator and altimeter is supplied by the fuselage-mounted pitot head. Static pressure is supplied by ports on each side of the fuselage forward of the cockpit area. The ball nose (1, figure 1-1) is a sphere-shaped, pitot-pressure, flow-direction sensor. The ball simultaneously measures angle of attack and angle of sideslip through two complete and independent servo systems. One system controls the vertical and one the horizontal axis. As the airplane encounters a sideslip condition or change in angle of attack, the respective system will turn the ball (electrohydraulically) into the relative wind. The difference between airplane heading and relative wind is then transmitted through sensors to cockpit instruments to read angle of attack and angle of sideslip. Electrically, the ball nose is powered through the instrumentation transformer, which in turn is powered by the No. 2 primary ac bus. Hydraulic power for the ball nose is provided from the No. 1 hydraulic system. If hydraulic system No. 1 fails, the ball nose will not turn in response to sensor signals. In this event, the angle-of-attack and sideslip indicators, although inoperative, will provide erroneous indications, as they will continue to register the conditions at the time of the power failure. In case of electrical failure, the indicators may show a continually unsafe condition.

Ball Nose Test Button. This push-button type switch on the center pedestal (figure 1-5) is labeled "BALL NOSE TEST" and is used to test operation of the ball nose in flight. Depressing and holding the button applies an error signal to the ball nose transducers. This error signal electrically simulates a predetermined airplane sideslip angle and angle of attack, which are presented on the angle-of-attack indicator and vertical and horizontal pointers on the attitude indicator. When the button is released, the ball nose should drive rapidly to an extreme position, resulting in full-scale deflection of the angle-of-attack indicator pointer and vertical and horizontal pointers on the attitude indicator. This indication should be maintained for about 2 to 3 seconds; then the ball nose should drive rapidly without overshoot to indicate the actual airplane sideslip angle and angle of attack.

NOTE

Full-scale deflection of the angle-of-attack indicator pointer and vertical and horizontal pointers on the attitude indicator when the button is released and the subsequent return to normal readings, are positive indications of proper operation of the ball nose.

ACCELEROMETER.

A three-pointer accelerometer (5, figure 1-2), on the instrument panel, shows positive and negative G-loads.

One pointer continuously indicates acceleration forces. Two recording pointers indicate maximum positive and negative G encountered. The recording pointers may be reset by clockwise movement of the reset knob on the lower left corner of the instrument.

ALTIMETER.

The altimeter (2, figure 1-2), on the instrument panel, has standard 1000- and 100-foot pointers and a 10,000-foot pointer extending from a movable center disk to the edge of the dial, so that it cannot be obscured by the other pointers. The center disk also has a wedge-shaped cutout through which a set of warning strips appear at altitudes below 16,000 feet. This altimeter offers improved readability and gives warning when an altitude of less than 16,000 feet is entered.

AIRSPEED INDICATOR.

The airspeed indicator (7, figure 1-2), on the instrument panel, shows indicated airspeed within a range of 100 to 900 knots with a conventional-type pointer. Visible through a window on the face of the indicator is a vernier drum which has a range of 0 to 100 knots. This permits reading of airspeed to the nearest knot through a range of 0 to 1000 knots. The airspeed indicator is the primary flight instrument for indicating speed during landing.

AIRSPEED/MACH INDICATOR.

The airspeed/Mach indicator (3, figure 1-2) has an airspeed needle fixed to a rotating plate with a range of 80 to 850 knots. The needle also indicates Mach number when airspeed is above 250 knots. The Mach scale rotates under the airspeed pointer plate and is visible through a cutout in the plate at airspeeds above 250 knots. The Mach scale rotates with altitude changes so that the indicating pointer will show the Mach number that equals the indicated airspeed for the particular flight altitude. The indicator has an adjustable airspeed setting index marker that is moved to the desired position by a knurled knob at the lower right corner of the instrument ring. The index marker can be used for referencing any desired in-flight speed as well as landing speed. A movable red hand is set by maintenance personnel to indicate 700 knots EAS. As altitude is increased, changes in outside air density cause the hand to move to a higher IAS reading.

ANGLE-OF-ATTACK INDICATOR.

A remote-type angle-of-attack indicator (4, figure 1-2), on the instrument panel, is electrically driven by power from both the No. 2 primary ac and primary dc busses. The ball nose measures the angle between the relative wind and the fuselage reference line. The attack angle so determined is then transmitted to the indicator. The indicator has a range from 10 degrees nose down to 40 degrees nose up.

RATE-OF-ROLL INDICATOR.

A rate-of-roll indicator (48, figure 1-2), on the instrument panel, is electrically powered by the primary dc bus. The rate-of-roll indicator indicates the roll rate in degrees per second for right and left roll from 0 to 200 degrees per second.

INERTIAL ALL-ATTITUDE FLIGHT DATA SYSTEM (GYRO-STABILIZED PLATFORM).

NOTE

For convenience of presentation, the inertial all-attitude flight data system henceforth is referred to as the stable platform system.

The stable platform system is essentially a navigating system designed to function over the earth within a high range area, approximately 720 miles long and 240 miles wide. Primarily, the system provides an attitude, velocity, and height record in a flight environment in which conventional flight reference instruments cease to function (that is, during prolonged operation at high altitudes and Mach numbers). The equipment in this system is operated with no angular range limitations or tumbling effects inherent in conventional gyro equipment. The system displays primary flight data to the X-15 pilot and transmits this data to recorders in the X-15 Airplane. The system is divided functionally into two groups. The first group, in the X-15, consists of the platform, computer, and flight instruments, which present attitude, velocity, and altitude indications. This equipment supplies all the required data after the X-15 is launched. The platform itself has four independent gimbals that permit unlimited maneuvers about the longitudinal, lateral, and vertical axes. It also has three gyros and three accelerometers. The second group, in the carrier airplane, is used to supply the proper initial conditions to the computer and thus align and stabilize the platform before launch. A control panel in the carrier airplane enables an operator to continuously monitor performance of the flight data system before launch, and to preset the required calibrations to the X-15 through an umbilical cord. In captive flight, the system is slaved to compass and velocity-measuring equipment in the carrier airplane. During free flight, the system dead-reckons from the launch point.

STABLE PLATFORM SYSTEM CONTROLS AND INDICATORS.

NOTE

In referring to electrical power sources in the following paragraphs, it is considered that the stable platform switch is positioned at INT and that the system controls and indicators are being powered from the X-15 Airplane.

Stable Platform Power Switch.

The stable platform power switch (26, figure 1-2) provides a means for selecting stable platform power, either from the carrier airplane or from the X-15. The switch is on the instrument panel right wing and is labeled "STABLE PLATF PWR." It has three maintained positions: INT, EXT, and OFF. The OFF position is a detent position. To move the switch from OFF, the switch must be pulled out of the detent. Moving the switch to INT energizes the platform with No. 2 primary ac bus power from the APU's of the X-15.

NOTE

Do not turn the stable platform switch to INT when the APU's are off.

When the switch is moved to EXT, the system is energized by power units in the carrier airplane.

NOTE

The switch should be moved from EXT to INT just before launch. However, if this is not done, power will be automatically transferred from the carrier airplane to the X-15 at time of launch.

Turning the switch to OFF shuts off all power to the platform system.

Stable Platform Instrument Switch.

This two-position switch on the center pedestal (figure 1-5) is labeled "S/P INST." The switch controls power to the stable platform indicators in the cockpit. It is used during ground checks of the stable platform system to interrupt power to the instruments when their operation is not required to perform the check-out. With the switch at OFF, power to the stable platform instruments is interrupted. Except during ground switch must be at on.

Attitude Indicator.

The attitude indicator (6, figure 1-2), on the instrument panel, is powered by the primary dc bus and the No. 2 primary ac bus. It is a pictorial-type instrument that combines displays of attitude and azimuth on a universally mounted sphere displayed as the background for a miniature reference airplane. The sphere (remotely controlled by the stable platform) is free to rotate 360 degrees in pitch, roll, and azimuth. The miniature reference airplane is always in proper physical relationship to the simulated earth, horizon, and sky areas of the background sphere. The horizon on the sphere is represented as a solid white line. On this horizon line is an azimuth scale graduated in 5-degree markings from 0 through 360 degrees. Above the horizon line, the sky is indicated by a light-gray area. Below the horizon line, the earth is indicated by a dull-black area. The sphere is marked by meridian lines spaced every 30 degrees. Pitch angle is referenced to the center dot of the fixed miniature airplane by horizontal marks spaced every 10 degrees on the meridians. A pitch-adjustment knob on the lower right side of the instrument electrically rotates the sphere to the proper position in relation to the miniature airplane to correct for pitch attitude changes. Clockwise rotation of the knob causes the horizon line to deflect upward from the airplane index. Rotating the knob counterclockwise causes the horizon line to deflect downward. Trim setting is automatically and gradually cancelled as airplane attitude approaches the vertical in climb or dive to ensure a true vertical indication. It returns automatically when level flight is resumed. Bank angles are read from a semicircular bank scale on the lower quarter of the instrument. Two long pointers project across the sphere. Movement of these pointers shows airplane displacement with respect to the air in which it is flying (small angles of attack and sideslip). The horizontal long pointer is a vernier indication of the angle-of-attack slip indicator on the instrument panel. This horizontal pointer moves upward when the angle of attack is increased and downward when the angle of attack is decreased. The vertical long pointer moves to the right to indicate a left sideslip, and to the left to indicate a right sideslip. The range of either pointer movement is adjustable (on the ground) to operate within ±5 to ±10 degrees. A short horizontal pointer is on the left side of the instrument. This pitch pointer is a vernier indication for the pitch axis of the sphere and moves in the same direction as the sphere. The pointer indicates displacement in a range of ±5 degrees of that angle selected on a pitch angle set control. Immediately after launch, the pilot rotates the airplane to the pitch angle that is preset on the pitch angle set control. Initially, the sphere is used to approximate this angle. As the airplane approaches to within 5 degrees of the preselected pitch angle, the small pitch pointer moves toward the center index (0). The pilot then switches his attention from the sphere to the pointer for fine adjustments. The pointer will remain at the zero position as long as the preset pitch angle is maintained. A turn-and-slip indicator is on the lower portion of the instrument below the bank angle scale. The rate-of-turn needle is powered by the 26-volt ac bus and the No. 2 primary ac bus. Immediately after electrical power is applied, an "OFF" flag on the lower left section of the indicator retracts. Failure of either dc or ac power causes the "OFF" flag to reappear.

Pitch-angle Set Control.

A pitch-angle set control (8, figure 1-2), powered by the primary dc bus and the No. 1 primary ac bus, is on the instrument panel, next to the attitude indicator. The control is used in conjunction with the small pitch pointer on the attitude indicator. The angle that is set on this control is the pitch angle the pilot will attain for either the climb or the re-entry phase of a mission. The instrument consists of four counters and a pitch angle set controller knob and lever. Rotating the knob clockwise sets up the desired pitch angle on three of the counters. The number on the far right counter is preceded by a dot to indicate the reading is in tenths of a degree. Counter range is from 0 to 90 to permit selection of any pitch angle up to 90 degrees. Rotating the knob counterclockwise returns the three counters to 0. The lever, adjacent to the knob, can be rotated upward or downward to change the sign (negative or positive) of the selected pitch angle. When the lever is moved upward, a minus (-) sign shows on the left counter; downward movement produces a plus (+) sign.

Section I

(Deleted)

Inertial Height Indicator.

The inertial height indicator (10, figure 1-2), powered by the primary dc bus and No. 1 primary ac bus, is on the instrument panel. The indicator is coupled to the stable platform system computer to automatically indicate any change in height. The computer, in turn, determines the height from vertical accelerations measured by a vertical accelerometer in the stable platform. Height is measured from an arbitrary reference line. This reference line may be determined in one of several ways - pressure altitude, ground control radar, etc. The inertial height indicator has a long 10,000-foot pointer and a short 100,000-foot pointer. Before launch, the pilot should readjust the height indicator (by turning a knob labeled "SET" on the face of the instrument) to his height in relation to the arbitrary reference line to eliminate possible cumulative errors.

Inertial Speed (Velocity) Indicator.

The speed (velocity) indicator (9, figure 1-2) is a single-pointer instrument coupled to the stable platform system computer to indicate airplane trajectory velocity. The computer determines this trajectory velocity from accelerations measured by accelerometers in the stable platform. The indicator, powered by the primary dc bus and No. 1 primary ac bus, is on the instrument panel and reads in thousands of feet per second from 0 to 7.

Vertical Velocity Indicator.

This single-pointer indicator (13, figure 1-2), on the instrument panel, displays inertial ascent and descent in hundreds of feet per second from 0 to 10. The vertical velocity indicator, powered by the primary dc bus and the No. 1 primary ac bus, is coupled to the stable platform system computer. The computer determines this vertical velocity from accelerations measured by the vertical velocity accelerometer in the stable platform.

INSTRUMENTATION SYSTEM

The instrumentation system records a wide variety of data on the basic airframe and airplane systems. The system is powered from the No. 2 primary ac bus and the primary dc bus. Controls and indicators for the system are on the center pedestal. (See figure 1-5.)

INSTRUMENTATION SYSTEM CONTROLS AND INDICATORS.

Instrumentation Master Power Switch.

Turning the two-position "INST MASTER" switch to on energizes all instrumentation heater circuits, energizes air borne recording instrumentation dc power circuits, and arms the gyro cage switch, strain gage power switch, and data switch. Moving the switch to OFF electrically de-energizes all instrumentation equipment.

Strain Gage Power Switch.

When the two-position "STRAIN GAGE" power switch is turned from OFF to ON, all transducers in the instrumentation system requiring strain gage battery power and that record on air-borne oscillographs, are energized.

Data Switch and Light.

This two-position "DATA" switch energizes all airborne recording media and arms the camera and calibrate switches when it is turned from OFF to ON. The switch also turns on a neon data light next to the switch. The data light blinks in synchronization with the camera timer, thus providing a positive indication of correct timer operation and energizing of recorder film magazines.

Calibrate Button and Light.

This is a spring-loaded, push-button type switch, labeled "CALIB," with an integral signal light. Momentarily depressing the button triggers the automatic in-flight calibration circuits used with the strain gage transducers to record the air-borne oscillographs. The signal light in the button comes on green when the button is depressed, to indicate that automatic calibration is in progress. The light continues to glow after the button is released, remaining on until the calibration is completed.

Camera Switch.

The "CAMERA" switch provides a means of selecting cine or pulse operation for the recording cameras. The switch has three positions; CINE, PULSE, and OFF.

Telemeter Master Power Switch.

This two-position switch, when turned from OFF to ON, energizes all power circuits in the telemeter system.

Telemeter Commutator Motor Switch.

Moving this two-position switch from OFF to COMM energizes the commutator motor.

Engine Oscillograph Record Switch.

The engine oscillograph record switch is on the center pedestal instrumentation panel (5, figure 1-5) and is labeled "ENG OSC." With instrumentation system electrical power available (instrumentation master and data switches ON), moving the engine oscillograph record switch to ON (up) applies primary dc bus power to operate the engine oscillograph. When the switch is at OFF, the oscillograph is off and will not record system parameters. Both switch positions are maintained. Use of the switch depends on the type of research mission to be flown.

Engine Vibration Recorder Switch.

The engine vibration recorder switch, on the instrumentation panel (5, figure 1-5), is labeled "VIB REC."

Moving the switch to ON (up) turns on the engine vibration recorder. Although a vibration sensor on the engine continuously sends engine vibration signals through a signal box to the recorder, no record is made until the engine vibration recorder switch is moved to the ON position. The recorder, which has its own battery power, is shut down when the switch is moved to OFF.

Physiological Instrumentation Switch.

A physiological instrumentation switch is on the instrumentation control panel (5, figure 1-5), on the center pedestal, on modified airplanes. The switch, labeled "PHYS INST," allows the physiological instrumentation system to be turned ON (switch moved up) or OFF. The switch controls primary dc bus power to the physiological instrumentation system.

INDICATOR, CAUTION, AND WARNING LIGHT SYSTEM.

Malfunctions or operating conditions for various airplane systems are indicated by placard-type indicator, caution, and warning lights on the instrument panel and center pedestal. For information on the functions of these lights, refer to the applicable systems. Except for the engine compartment fire-warning light, all the lights can be tested by a switch in the cockpit.

INDICATOR, CAUTION, AND WARNING LIGHT SWITCH.

An indicator, caution, and warning light switch (31, figure 1-2), powered by the primary dc bus, is on the instrument panel right wing. This switch has two positions, TEST and NORMAL. When the switch is placed in the TEST position, all indicator, caution, and warning lights (except the fire-warning light) on the instrument panel and center pedestal come on; this is only a test of the bulbs. The switch must be at NORMAL for all normal operations.

CANOPY.

The one-piece, pneumatically counterbalanced, clamshell-type canopy (figure 1-13) is manually operated and mechanically locked in the down position. The canopy is hinged at the rear and opens about 45 degrees after moving aft to unlock. The canopy can be manually locked from either inside or outside. The canopy has a double-pane window with an air space between the panes for defrosting air. A retractable, adjustable head support can be lowered from the top of the canopy to restrain the pilot's head during deceleration. A canopy seal, incorporated in the rim of the canopy, contacts the canopy sill and a bulkhead at the rear of the cockpit to allow pressurization of the cockpit. A cartridge-type canopy remover is fired by an initiator when the ejection seat armrests are raised. Also, this initiator directs some of its expanding gases to extend a thruster at the forward end of the canopy. When the canopy leaves the airplane, it fires the ejection seat catapult initiator to eject the seat. The canopy can also be ejected with the canopy internal emergency jettison "T" handle which fires a separate initiator that does not arm and fire the ejection seat when the canopy leaves. A similar "T" handle is provided externally, behind a door on the right side of the fuselage, just below the canopy split line, for ground emergencies.

CANOPY SEAL.

An inflatable rubber seal, built into the edge of the canopy frame, seats against mating surfaces of the canopy sill and a bulkhead at the aft end of the cockpit to provide sealing for cockpit pressurization. The seal pressurization valve is mechanically actuated just before the complete locking of the canopy to permit gaseous nitrogen to inflate the seal. When the canopy handle is actuated to open the canopy, the seal pressurization is dumped and the nitrogen valve is closed.

CANOPY CONTROLS.

Canopy External Handle.

The canopy external handle (figure 1-13), on the right side of the fuselage, below the forward end of the canopy, is behind a flush door that is opened by pushing a flush-mounted button just aft of the door. Pulling the black long-hinged handle out of its spring clip and rotating it upward unlocks and moves the canopy aft about one inch to permit manual raising of the canopy. Rotating the handle forward and down moves the canopy forward and locks it. Pushing the handle into the spring clip stows the handle.

Canopy Internal Handle.

The canopy internal handle (1, figure 1-4), for locking or unlocking the canopy, is on the right side of the cockpit, just below the canopy sill. Locking of the canopy requires that the canopy be lowered manually until it is tight against the canopy sill, then the handle pushed forward until it is against the stops and then rotated outboard to lock. Unlocking the canopy is accomplished by pulling the handle inboard and aft all the way back against the stops until the canopy is unlocked. The canopy can then be raised manually. During the ejection sequence, the handle automatically moves aft when the canopy unlocks.

Keep hands and arms clear of canopy internal handle when canopy is jettisoned, because the canopy handle moves aft with considerable force.

Canopy Internal Emergency Jettison Handle.

The canopy internal emergency jettison handle (25, figure 1-2), on the instrument panel right wing, jettisons the canopy without firing the seat. The "T" handle fires a separate initiator that fires the canopy remover and

CANOPY

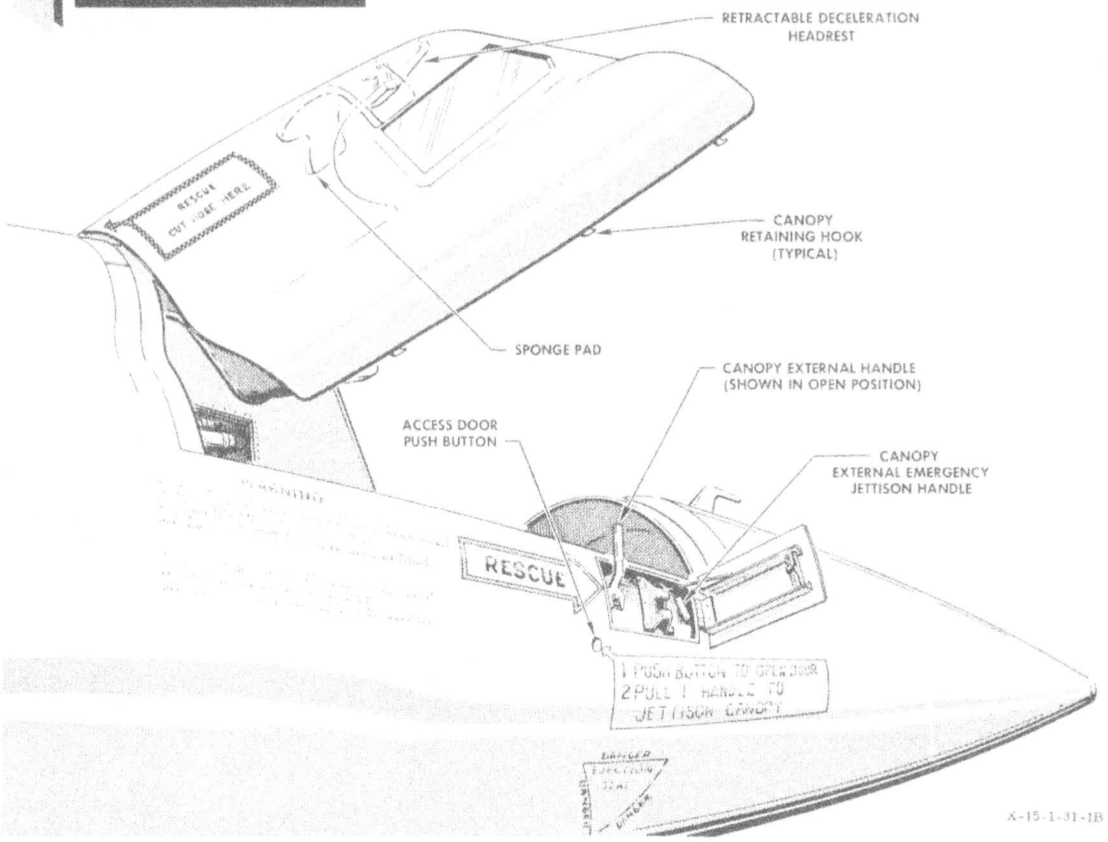

Figure 1-13

extends the canopy thruster. On the ground, the handle is safetied with a safety pin inserted through the handle.

Keep hands and arms clear of canopy internal handle when pulling the emergency jettison handle, because the canopy internal handle moves aft with considerable force.

Canopy External Emergency Jettison Handle.

The canopy may be jettisoned from the outside by the canopy external emergency jettison handle. (See figure 1-13.) This yellow, T-type handle is just forward of the canopy external handle in a recess behind a flush door on the right side of the fuselage, below the forward end of the canopy. When the handle is pulled straight out approximately 4 inches with a force of about 10 pounds, a canopy initiator is fired. This in turn fires a canopy remover that forcibly jettisons the canopy.

The seat is not armed when the canopy is removed by this means.

- The canopy external emergency jettison handle does not have an extension lanyard. Therefore, extreme care should be taken to ensure that no part of a person's body is directly over any portion of the canopy when it is jettisoned.

- If the ejection handles are raised and the canopy has not jettisoned, the ejection seat must be deactivated before the canopy is either manually opened or jettisoned, to prevent seat ejection.

SAFETY PINS.

Because of the interdependence of the seat and canopy ejection systems, the seat and canopy safety pins are discussed together in this paragraph. To safety the canopy and seat ejection systems adequately, seven safety pins are required. The two initiators on the canopy deck, just aft of the right side of the seat, and the two initiators on the right side of the seat near the headrest are safetied by a safety pin inserted through a hole in the initiator sear pin of each initiator. The right-hand ejection handle, on the ejection seat, restraint emergency release handle on the ejection seat, and canopy internal emergency jettison handle on the instrument panel right wing are safetied with a safety pin through a hole in each handle.

WARNING

All safety pins must be removed before flight and replaced after landing.

EJECTION SEAT.

The ejection seat (figure 1-14) is designed to permit safe pilot ejection up to Mach 4.0, in any attitude, and at any altitude up to 120,000 feet. Firing of the seat is initiated by jettisoning the canopy. The seat cannot be ejected unless the canopy has left the airplane. A ballistic-rocket type catapult supplies the necessary propulsion force to eject the seat and pilot from the airplane. During ejection, stabilizing fins and booms automatically extend to stabilize the seat. Restraint devices are provided for the pilot's body and legs to prevent injuries and separation from the seat above 15,000 feet. At this altitude, an aneroid device fires three initiators to free the restraint devices and permit pilot separation from the seat. A manual handle is provided to permit the pilot to release the restraints if the aneroid device fails. If ejection below 15,000 feet is accomplished, there is a 3-second delay after ejection before automatic pilot-seat separation is initiated. The breathing oxygen supply is contained in cylinders mounted to the underside of the seat and is used when the airplane is launched from the carrier airplane. While in captive flight, breathing oxygen is supplied by the carrier airplane. An oxygen selector valve and gage on the left side of the seat permits selection of either carrier airplane oxygen or the seat-contained oxygen. The personal leads (radio, face mask heater, oxygen, and ventilated suit) are attached to a disconnect block that is fitted into a disconnect fitting on the left side inside the seat bucket. A battery, mounted under the right stabilizing fin, automatically supplies power for face mask heating during postejection. The ejection seat also has a quick-disconnect receptacle to plug in the pilot's physiological instrumentation wiring harness. The receptacle is on the top of the seat pan in the forward right-hand corner. The manually adjustable shoulder harness straps are fastened to the integrated parachute restraint harness with quick-disconnect fittings. During the ejection sequence, the shoulder harness is released when the headrest ejects. The pilot's parachute is carried in a container attached to the pilot's integrated harness, with a pilot chute in a separate container; the pilot chute is released when the headrest is ejected. The operation sequence of the aneroid device is actuated at 15,000 feet, and it fires the three initiators that fire the headrest and release the seat belt, personal leads, ejection handles, and all restraints, to permit pilot separation from the seat.

EJECTION SEAT CONTROLS.

Ejection Seat Ejection Handles.

Unlatching and raising either ejection handle on the seat to within 15 degrees of its full travel fires an initiator that fires the canopy remover. As the canopy leaves the airplane, it fires the seat catapult initiator that fires the two-stage seat catapult. The ejection handle release latch is in the top portion of each ejection handle and is actuated when either ejection handle is grasped. Since the ejection handle assemblies are linked together by a linkage to a torque tube, pulling up on either ejection handle automatically raises the other. The ejection handles lock in the full travel position until unlocked by the restraint release system. If the canopy is inadvertently lost in flight, the seat will not eject unless the ejection handles are raised. During pilot separation from the seat after ejection, the ejection handle assemblies are automatically unlocked and swing out board to permit unrestricted pilot separation from the seat. Modified ejection seats and pressure suits provide automatic actuation of the pilot's emergency oxygen supply when the ejection handles are raised. As the handles are raised, a cable pulls the pin from the emergency oxygen supply valve. The cable is attached in parallel with the manual actuation cable.

Restraint Emergency Release Handle.

A restraint emergency release handle, on the right side of the ejection seat, is pulled up to afford a quick release from the seat on the ground when the manual release would be too slow, or after ejection if the aneroid device fails to actuate the automatic restraint release system. Pulling this handle releases the foot restraints, lap belt, personal leads, and armrest assemblies, and fires the headrest which releases the shoulder harness.

Foot Restraint Release Buttons.

The foot restraint release buttons are on the top front corners of the seat, above each foot restraint. Depressing each button unlocks its respective foot restraint, which releases the pilot's feet. During automatic pilot-seat separation, the foot restraints unlock automatically.

SAFETY PINS.

Refer to "Safety Pins" under "Canopy" in this section.

PARACHUTE, INTEGRATED RESTRAINT EQUIPMENT, AND PRESSURE SUIT.

The personal parachute for the pilot is contained within a fiber glass parachute container that is attached to the pilot's integrated harness at the top and bottom. The

Section I T.O. 1X-15-1

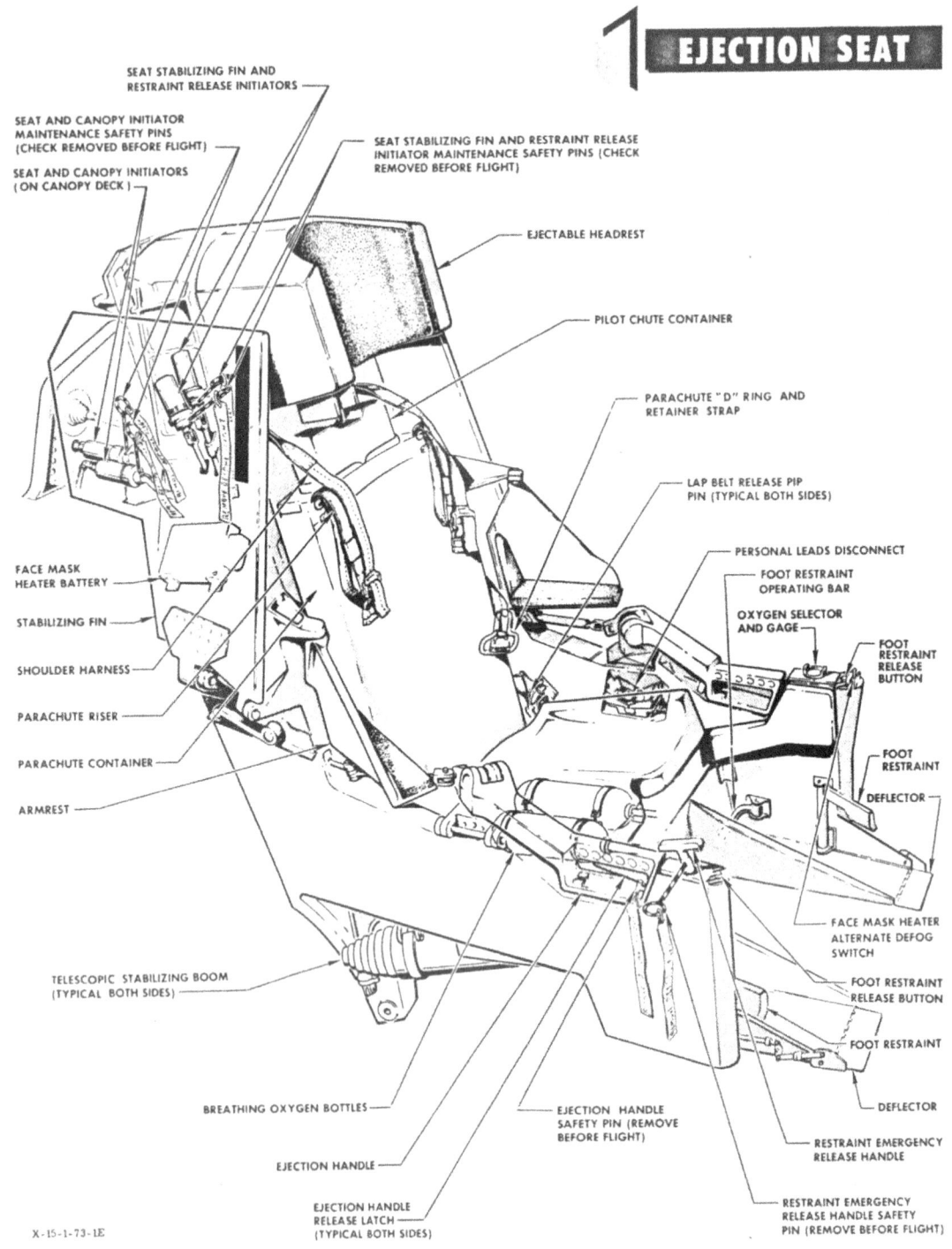

Figure 1-14

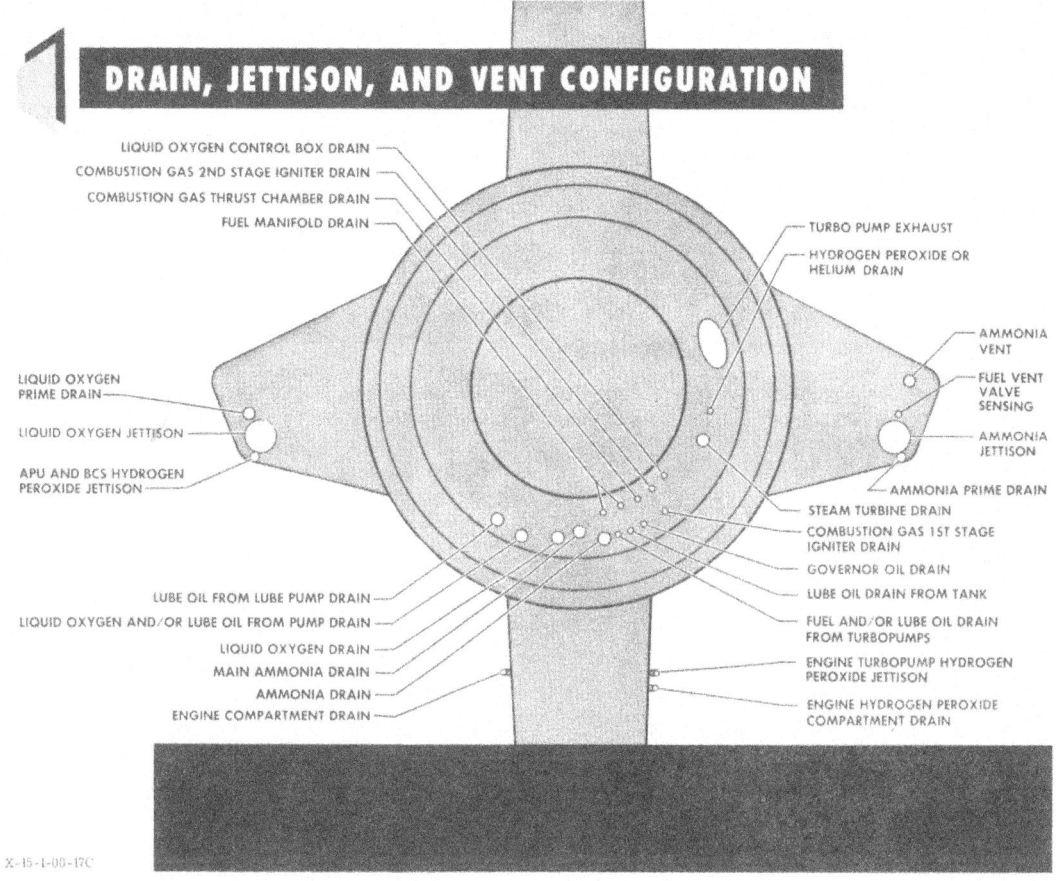

Figure 1-15

parachute may be either a 24-foot or a 28-foot type. Another fiber glass container contains the pilot chute and, along with the parachute container lids, closes the top of the parachute container. The pilot chute container is held in place by a retaining pin that locks into the manual ripcord pin. On the top of the container, a retainer pin fits into the ejection seat headrest. During ejection, the pilot chute container is locked to the headrest as the seat leaves the airplane, and is pulled off the parachute container to deploy the pilot chute when the headrest is fired. This in turn pulls out the main parachute canopy. A rescue beacon transmitter, installed in the parachute container, is automatically energized into continuous operation when the pilot's parachute deploys. The transmitter antenna is attached to one of the parachute straps. Transmission is on the X-15 telemetering frequency of 244.3 megacycles. The transmitter permits ground stations to obtain position fixes on the pilot after an ejection. The parachute container is attached to the seat by a strap on the lower corner of each side. This strap is attached to a fitting which is held by a lap belt release on each side in the seat bucket. This release hinges on a pip pin during automatic release, but can be removed manually to permit removal of the parachute container without actuating the restraint release system. When pilot-seat separation occurs during ejection, the parachute container remains with the pilot. The parachute riser quick-release buckles also fasten the shoulder harness and parachute container to the pressure suit fittings, on the pilot's chest, just below the shoulders. The rip cord "D" ring is attached to a strap fastened to the left side of the parachute container. This strap also has a quick-release buckle which is fastened to a fitting on the left side of the integrated restraint harness, just below the arm. The lap belt portion of the integrated restraint harness consists of two straps, one over each hip, fitted with hooks that are snapped into a ring portion of the lap belt release fitting on either side of the seat. The lap belt has double adjuster straps, one through each side of the buckle. The adjuster straps must be pulled tight to keep the pilot firmly in the seat. This attaches the pilot to the seat and parachute container. The full pressure suit was modified for the X-15 Airplane and has the restraint straps and parachute harness designed as an integral part of the suit. A neck seal is used to keep the suit pressurization nitrogen and breathing oxygen separated. Attached to the back of the restraint harness is a controller back pan which incorporates an oxygen regulator, suit pressure regulator, anti-G valve, and the emergency oxygen supply for breathing and suit emergency pressurization. This emergency oxygen supply is sufficient for about 20 minutes after the pilot separates from the ejection seat. The oxygen and communication lines from the controller back pan are internal and plug into a pressure suit mating receptacle over the left shoulder blade. The personal-lead disconnect block is also attached to the controller back pan. The suit regulator ventilation exhaust valve is over the right shoulder blade. The emergency oxygen supply in the controller back pan is actuated automatically (modified ejection seats and pressure suits) when the seat ejection handles are raised, or can be actuated manually by a green ball on the right side of the suit. The helmet is free-rotating and is fastened to the suit with a snap connector ring that seals the joint. The helmet visor is locked down with a squeeze latch on the bottom edge.

WARNING

If the pilot prebreathes 100 percent oxygen before entering the cockpit, the visor must not be opened at any time during ground or flight operations; otherwise, the denitrogenation effect will be nullified.

When the visor is down and latched, the suit and helmet are completely sealed in a leakproof unit. The visor is kept fog-free by the breathing oxygen as it comes in around the face opening and across the visor.

AUXILIARY EQUIPMENT.

The following auxiliary equipment is described in Section IV: air conditioning and pressurization system, communication and associated electronic equipment, lighting equipment, pilot's oxygen system, and miscellaneous equipment.

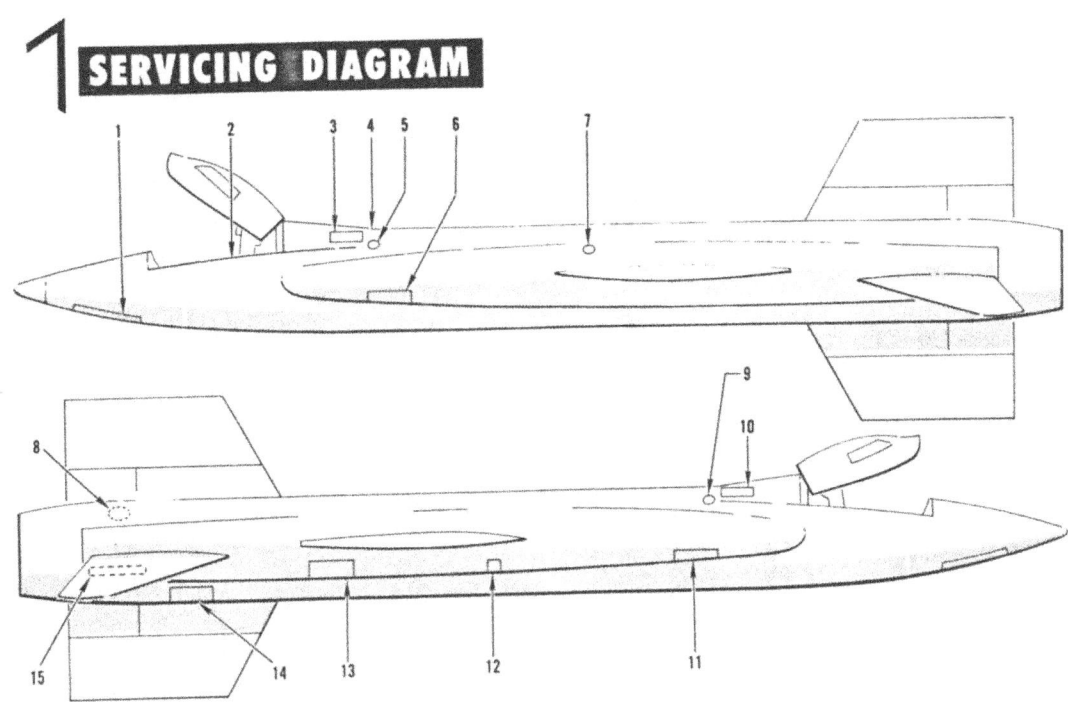

1 SERVICING DIAGRAM

1. Nitrogen filler—nose gear strut (in wheel well)
2. Breathing oxygen filler (on seat)
3. Lubricating oil filler—APU No. 1
4. B-52 disconnects (7)
5. No. 1 hydraulic reservoir level sight gage
6. Helium filler and pressure gage—APU No. 1
   Hydrogen peroxide filler—APU No. 1
   No. 1 hydraulic pressure disconnect
   No. 1 hydraulic suction disconnect
7. Liquid oxygen filler—engine oxidizer
8. Engine control hydraulic oil filler
9. No. 2 hydraulic reservoir level sight gage
10. Lubricating oil filler—APU No. 2
11. Liquid nitrogen filler—air conditioning and pressurization system
    Helium filler—air conditioning and pressurization system
    Helium filler and pressure gage—APU No. 2
    Hydrogen peroxide filler—APU No. 2
    No. 2 hydraulic pressure disconnect
    No. 2 hydraulic suction disconnect
12. Liquid nitrogen filler—helium tank cooling
    Pressure test—propellant system controls
    Helium filler—propellant system
13. Ammonia filler
14. Helium filler—engine purge system and No. 1 and No. 2 hydraulic accumulators
    Hydrogen peroxide filler—engine turbopump propellant
15. Engine lubrication system filler

SPECIFICATIONS

HYDROGEN PEROXIDE	NA2-2103 Grade A
HELIUM	High-grade, oil-free, with purity greater than 99.9 percent, and maximum dew point of $-70°F$
GASEOUS NITROGEN	Grade A Type I MIL-N-6011
LIQUID NITROGEN	Grade A Type II MIL-N-6011
LIQUID OXYGEN (OXIDIZER)	MIL-P-25508
GASEOUS OXYGEN (BREATHING)	MIL-O-27210
HYDRAULIC FLUID	NA2-2078A (Oronite 8515)
LUBRICATING OIL (APU)	MIL-L-7808C
LUBRICATING OIL (ENGINE)	Halo carbon oil 4-11V (RMD Spec 4043)
ANHYDROUS AMMONIA	JAN-A-182
HYDRAULIC OIL (ENGINE CONTROL)	RMD Spec 4041

Figure 1-16

THIS PAGE INTENTIONALLY LEFT BLANK.

Normal Procedures

SECTION II

TABLE OF CONTENTS	PAGE NO.
Introduction	2-1
Preparation for Flight	2-1
Preflight Check	2-1
Captive Taxi and Flight	2-6
Prelaunch	2-7
Ballistic Control and Reaction Augmentation System Operation	2-12
Launch	2-12
Flight Characteristics	2-12
Engine Operation	2-12
Engine Burnout	2-12
Shutdown Procedure	2-14
Aborted Launch	2-14
Descent	2-14
Before Landing	2-14
Landing	2-14
After Landing	2-14
Before Leaving Airplane	2-15

INTRODUCTION.

The research mission of the X-15 requires flexibility of operation; for this reason, the procedures presented in Section II are subject to change. The procedures presented are for a typical mission, and do not reflect any special research requirements.

PREPARATION FOR FLIGHT.

PILOT-COCKPIT COMPATIBILITY.

For efficient flying of the airplane and to ensure optimum escape conditions if ejection is necessary, the pilot-cockpit compatibility adjustments must be accomplished as outlined in the X-15 Maintenance Manual, Report No. NA-58-770.

FLIGHT RESTRICTIONS.

Refer to Section V for detailed airplane and engine limitations.

WEIGHT AND BALANCE.

Refer to Section V for weight and balance limitations. For loading information, refer to Weight and Balance Technical Manual, T.O. 1-1B-40.

ENTERING COCKPIT.

When the airplane is in place on the pylon of the carrier airplane, the cockpit can be entered from either side.

A hydraulically operated, adjustable stand is raised to the cockpit ledge to permit normal entry to the cockpit. (See figure 2-1.)

If the pilot prebreathes 100 percent oxygen, the visor must not be opened at any time during ground or flight operation.

PREFLIGHT CHECK.

BEFORE EXTERIOR INSPECTION.

Check Form 781 for engineering status and to make sure that the airplane has been serviced for the scheduled mission. (See figure 1-16 for servicing points.)

EXTERIOR INSPECTION.

Because of the mission of this airplane and the personal equipment used by the pilot, it is not feasible for the pilot to perform an exterior inspection.

CANOPY AND EJECTION SEAT CHECK.

Before entering cockpit, check canopy and ejection seat as follows:

1. Canopy ejection mechanism - Check.
    Open canopy fully to visually check canopy remover mechanism and that initiator ground safety pins are

1 ENTERING COCKPIT

Figure 2-1

installed. Also check tubing and fitting from canopy initiators to canopy remover.

2. Foot restraints - Check visually.

   Check that foot restraints are in the open position.

3. Ejection seat safety pins - Check visually.

   Check that ejection seat safety pins are installed in seat initiators. Also check tubing and fittings from initiators to seat ejection catapult.

4. Canopy external handle and external emergency jettison handle - Check.

   Make sure canopy external manual and emergency jettison handles are stowed properly and that door is closed.

5. Ejection handles - Down and latched.

INTERIOR CHECK.

Before the interior check of the cockpit is made, certain instrument readings must be checked. (See figure 2-2.)

Personal Equipment.

Before the interior check of the cockpit is made, the following personal equipment should be connected and adjusted with the aid of ground personnel:

1. Pressure suit integrated harness - Attach harness to parachute risers and shoulder harness straps.

2. Integrated restraint harness - Fasten to seat.

3. Emergency oxygen system actuating cable - Connected.

   Check that the emergency oxygen system actuating cable is connected to the right-hand seat ejection handle.

4. Personal equipment quick-disconnect - Insert and lock.

COCKPIT INSTRUMENT READINGS AFTER SERVICING

NOTE
All readings are based on requirements for a full-duration mission.

INSTRUMENT		LOW	NORMAL	HIGH	REMARKS
PROPELLANT SOURCE PRESSURE GAGE		3500	3500 to 3600	3600	High limit is below minimum relief valve setting.
H_2O_2 SOURCE AND PURGE PRESSURE GAGE (BOTH POINTERS)		3500	3500 to 3600	3600	
APU SOURCE PRESSURE GAGE (BOTH POINTERS)		3600	3700 to 3900	3900	High limit is below minimum relief valve setting.
CABIN HELIUM SOURCE PRESSURE GAGE		3500	3500 to 3600	3600	
HYDRAULIC TEMPERATURE GAGE		−55°C	−30° to 50°C	95°C	During hydraulic servicing, hydraulic cart should be hooked up and fluid circulated to maintain temperature within limits shown. Hydraulic pressure should be maintained within limits shown.
HYDRAULIC PRESSURE GAGE (BOTH POINTERS)		3200	3400 to 3600	4500	
AC VOLTMETERS		190	190 to 210	210	Limits shown are required for proper operation of airplane components and are controlled by ground cart output.
PILOT'S OXYGEN PRESSURE GAGE		2950	2950 to 3300	3300	
H_2O_2 TANK AND ENGINE CONTROL LINE PRESSURE GAGE	"C" POINTER	575	575 to 600	620	High limit indicates excessive regulator leakage.
	"T" POINTER		0		Tank vented.

X-15-1-51-10

Figure 2-2

When the pressure suit quick-disconnect is inserted and locked, oxygen, communication, and ventilated suit leads are mated and connected.

5. X-15 LN_2 supply switch - Check OPEN.

   Confirm with launch operator that the switch is OPEN.

6. Rudder pedals.

7. Pilot's oxygen system selector - B-52.

7A. Face mask heater alternate switch - Check OFF.

8. Physiological package wiring - Connect.

After connecting and adjusting personal equipment, make a left-to-right check of the cockpit by sections as shown in the paragraphs that follow.

NOTE

Electrical power will be supplied from the carrier airplane at this time.

Left Console.

1. Pressure suit ventilation knob - As desired.

2. Vent suit heater switch - HIGH or LOW, as desired.

3. Radio controls - OFF.

4. Face mask heater switch - OFF.

5. Intercom switch - ON.

6. (Deleted)

7. Trim control switch - Normal.

8. Ready-to-launch switch - OFF.

9. Wing flap switch - UP.

10. Jettison trim switch - OFF (Center).

11. Speed brake handles - CLOSED (forward).

12. Ballistic control stick - Check attachment and freedom of movement.
13. RAS function switches - STANDBY.
14. RAS accelerometer switch - OFF.
14A. Alternate SAS switch - OFF.
14B. Hydrogen peroxide transfer switch - Check OFF.
15. Vent, pressurization, and jettison control lever - VENT.

The vent valve on the NH_3 tank will be manually closed before flight, to prevent losing NH_3 during captive flight. When the vent, pressurization, or jettison control lever is placed in the PRESSURIZE or JETTISON position and then back to VENT, the NH_3 vent valve will then be open.

16. Throttle - OFF.
16A. APU water cooling switches - Check OFF.
17. Jettison stop switches - STOP.

Check that all three switches (LOX, H_2O_2, and NH_3) are in the STOP position.

18. Auxiliary launch switch - OFF (guard down).
19. Landing gear handle - IN.

Instrument Panel (Engine Instruments).

1. Propellant emergency pressurization switch - OFF.
2. Ignition-ready light - Check OFF.
3. No-drop caution light - Check OFF.
4. Idle-end caution light - Check OFF.
5. Valve malfunction caution light - Check OFF.
6. Stage 2 ignition malfunction caution light - Check OFF.
7. Turbopump overspeed caution light - Check OFF.
8. Engine vibration malfunction caution light - Check.
9. Fire-warning light - Check OFF.
10. Helium release selector switch - OFF.
11. Engine master switch - OFF.
12. Engine reset button - Check.
13. Engine precool switch - OFF.
14. Engine prime switch - OFF.
15. Turbopump idle button - Check.
16. Igniter idle switch - OFF.
17. Propellant tank pressure gage - Check.

18. Propellant pump inlet pressure gage - Check.
19. Fuel line low caution light - Check.
20. Propellant manifold pressure gage - Check.
21. H_2O_2 compartment-hot caution light - Check OFF.
22. Chamber and stage 2 igniter pressure gage - Check.
23. Liquid oxygen bearing temperature gage - Check.

Instrument Panel (Flight Instruments).

1. Pilot's oxygen-low caution light - Check.
2. Accelerometer - Set and check.
3. (Deleted)
4. Altimeter - Set.
5. Attitude indicator - Set.
6. (Deleted)
7. Pitch angle set knob - As required.

Instrument Panel (Electrical, Hydraulic, and Cockpit).

1. Emergency battery switch - OFF (guard down).
2. No. 1 generator-out light - Check OFF.
3. No. 2 generator-out light - Check OFF.
4. No. 1 and No. 2 generator switches - OFF.
5. APU No. 1 switch - OFF.
6. APU No. 1 warning and caution lights - Check OFF.
7. No. 1 ballistic control switch - OFF.
8. No. 2 ballistic control switch - OFF.
9. APU No. 2 warning and caution lights - Check OFF.
10. APU No. 2 switch - OFF.
11. APU H_2O_2 pressure gage - Check 0 psi.
12. Clock - Check and set.
13. Mixing chamber temperature gage - Check.
14. APU bearing temperature gage - Check.
15. Cabin pressure altimeter - Check.

Center Pedestal.

1. Pitch function switch - STDBY.
2. Roll function switch - STDBY.

3. SAS test switch - Check OFF.

4. "Yar" function switch - STDBY.

5. Yaw function switch - STDBY.

6. SAS caution lights (four) - Check ON.

7. SAS gain selector knobs - Set.

    Set the gain selectors at the following: Pitch, 8; roll, 6; yaw, 8.

8. Ball nose test button - Check.

9. Research instrumentation - Set as required for mission.

10. Engine oscillograph record switch - As required.

11. Ram-air lever - CLOSED.

12. Radar beacon switch - As required.

13. Stable platform instrument switch - ON.

14. Engine vibration recorder switch - OFF.

15. Cockpit ram-air knob - OFF (in).

16. DC voltmeter selector switch - BUS.

17. DC voltmeter - Check.

Right Console.

1. Canopy emergency release handle - IN.

2. Stable platform power switch - EXT.

3. Nose ballistic rocket heater switch - OFF.

4. Ventral arming switch - DE-ARM.

5. Cockpit lighting switch - OFF.

6. Indicator, caution, and warning light switch - NORMAL.

7. Fire-warning light test button - Push to test.

    Fire-warning light ON indicates continuity of detection circuit.

8. Windshield heater switches (two) - OFF.

9. Circuit-breaker panel - Check circuit breakers as required.

10. Restraint emergency release handle - IN.

11. Pressure cooling lever - OFF.

12. No. 1 blower switch - OFF.

13. No. 2 blower switch - OFF.

14. Cabin source helium shutoff valve switch - OFF.

15. **APU nitrogen cooling switch - SINGLE.**

16. Alternate cabin pressurization switch - IN.

17. Console control stick - Check attachment and freedom of movement.

18. Center control stick - Check attachment and freedom of movement.

19. Alternate trim switch - Centered.

    Check alternate trim switch on the center control stick for freedom of operation.

20. Windshield purge handle - OFF (down).

21. Windshield anti-fogging handle - OFF (down).

Interior Inspection Operational Check.

1. Augmented cooling system - ON.

    Refer to "Operation of Ram-air and Augmented Cooling System (Captive Flight)" in Section IV.

2. Oxygen system - Check.

    Check that breathing oxygen is being supplied from the carrier airplane.

3. Face mask heater - Check.

    Move face mask heater switch to HI and check operation, then move switch to OFF.

4. Intercom - Check.

    Check communication between carrier airplane and X-15 Airplane.

5. DC voltmeter switch - BUS.

6. Calibrate instrumentation.

7. Launch light - Test.

    Push to test launch light; have verification from carrier pilot and launch panel operator.

8. Indicator, caution, and warning lights - Check.

    Place the indicator, caution, and warning light test switch at TEST. All indicator, caution, and warning lights (except the fire-warning light) will come on; this is only a test of the bulbs.

9. Ground safety pins - Removed.

    Have crew chief remove the six ground safety pins and display pins.

10. Close canopy.

COCKPIT INSTRUMENT READINGS BEFORE TAKE-OFF

NOTE
All readings are based on requirements for a full-duration mission.

INSTRUMENT		LOW	NORMAL	HIGH	REMARKS
PROPELLANT SOURCE PRESSURE GAGE		3300	3500 to 3600	3900	
H_2O_2 SOURCE AND PURGE PRESSURE GAGE (BOTH POINTERS)		3000	3500 to 3600	3900	Low limit is minimum pressure required for full-duration mission. High limit is maximum pressure relief valve setting.
APU SOURCE PRESSURE GAGE (BOTH POINTERS)		3300	3600 to 3900	4200	
CABIN HELIUM SOURCE PRESSURE GAGE		3300	3500 to 3600	3900	Relief valve set for 3900 maximum.
HYDRAULIC TEMPERATURE GAGES		−125 C	−75 C to 0 C	95 C	Temperature below low limit can cause excessive pressure drop and restriction of flow.
AC VOLTMETERS		190	190 to 210	210	Limits shown are required for proper operation of airplane components.
H_2O_2 TANK AND ENGINE CONTROL LINE PRESSURE GAGE	"C" POINTER	575	575 to 600	620	High limit indicates excessive regulator leakage.
	"T" POINTER		0		Tank vented.
PROPELLANT TANK PRESSURE GAGE	"A" POINTER	0	0 to 30	68	Vent valve manually closed.
	"L" POINTER		0 to 5		Tank vented.

X-15-1-51-11

Figure 2-3

CAPTIVE TAXI AND FLIGHT.

TAXI.

1. SAS function switches - Check.

   Move SAS function switches to ENGAGE and check; functions should trip during taxiing because of carrier airplane motions and no hydraulic pressure. Return function switches to STDBY after each function trips.

2. Radar beacon switch - ON.

BEFORE TAKE-OFF.

Before take-off of the carrier airplane, recheck the following:

1. Ram-air lever - CLOSED.
2. N_2 or helium release switch - AUTO.
3. Jettison stop switches - STOP.
4. Instrument Reading - Check.

   See figure 2-3 for proper instrument readings.

TAKE-OFF

During take-off, monitor all instruments, and relay to the carrier pilot or launch operator any information that could affect the planned mission.

1. Ventral arming switch - ARM.
2. Windshield heater switches - ON.
3. Engine master switch - ON.
4. Engine reset button - Push (one second).
5. Engine precool switch - PRECOOL.

   Precool the engine for 10 minutes on, then 20 minutes off, as instructed by ground control.

CLIMB.

After take-off, during the climb and cruise part of the flight, the liquid oxygen tanks of the X-15 Airplane will be topped off from the carrier airplane.

1. Instrument power switch - ON.

2. Telemetering power switch - ON.

3. Course indicator - Check operation.

4. Face mask heater switch - As required.

5. Communications - Check.

   Confirm radio communication with chase plane and ground station, and also that communication is available between X-15 Airplane, chase plane and carrier airplane, and chase plane and carrier airplane with ground station.

6. Liquid oxygen top-off cycle - Check.

   Confirm that liquid oxygen top-off from the carrier airplane liquid oxygen climb tank is satisfactory.

7. Hook heater - Confirm ON.

   Confirm with launch operator that hook heater is ON.

8. Nose ballistic rocket heater switch - ON.

9. Blower switches - OFF (15,000 feet).

   Move blower switches to OFF at 15,000 feet altitude.

10. Ram-air lever - OPEN.

CRUISE-CLIMB TO LAUNCH ALTITUDE.

During cruise-climb to launch altitude, the pilot of the carrier airplane will start the time-to-go sequence. Confirm with the launch operator that liquid oxygen top-off is satisfactory and that the liquid oxygen cruise tank has been started to complete the top-off.

At start of climb from 35,000 feet cruise altitude to 45,000 feet launch altitude, accomplish the following:

1. Ram-air lever - CLOSED.

2. Blower switches - BLOWER ONLY.

NOTE

If cabin pressure of 35,000 feet is not maintained during climb to 45,000 feet move blower switches to BLOWER & LN_2.

Engine Check.

Check and report the instrument readings. If instruments are not within limits, check with ground control for alternate mission if below limit.

NOTE

Values exceeding limits noted may indicate regulator failure.

PRELAUNCH.

BEFORE COUNTDOWN.

Before countdown, check location and time to go; then complete final cockpit check as follows:

1. Engine precool switch - OFF.

   After precooling is completed, return precool switch to OFF.

NOTE

If a delay in launch is required, the engine can be maintained in a precooled condition for an extended period of time by moving the engine precool switch to PRECOOL for approximately 7-1/2 minutes every 20 minutes.

2. Ram-air lever - CLOSED.

The oxygen supply to the pressure suit helmet must be on and the helmet visor must be down before the cockpit is pressurized with nitrogen.

3. Cabin source helium shutoff valve switch - CLOSED.

4. Pressure cooling lever - ON.

5. APU cooling switch - NORMAL.

6. Blower switches - BLOWER & LN_2.

7. Oxygen system - Switch from carrier supply to X-15 Airplane supply.

   Rotate oxygen selector valve on seat leg guard to X-15 Airplane. Check for oxygen flow.

8. X-15 LN_2 supply switch - Check CLOSED.

   Confirm with the launch operator that the switch is CLOSED.

2-7

Section II T.O. 1X-15-1

9. Inertial height indicator - Set.

10. Ventral arming switch - ARM.

11. Intercom switch - OFF.

12. Communications - Check.

   Check communications with ground station, carrier pilot, and chase pilot.

13. Intercom switch - ON.

14. APU No. 1 - Start.

   As APU No. 1 comes up to speed, hydraulic pressure will increase to as much as 4500 psi for a short time. However, pressure should stabilize at 3200 to 3400 psi within 30 seconds after APU No. 1 starts.

15. APU No. 2 - Start.

   As APU No. 2 comes up to speed, hydraulic pressure will increase to as much as 4500 psi for a short time. However, pressure should stabilize at 3200 to 3400 psi within 30 seconds after APU No. 2 starts.

16. No. 1 generator switch - ON.

   Move No. 1 generator switch momentarily to RESET, then to ON. Check No. 1 generator voltage.

17. No. 2 generator switch - ON.

   Move No. 2 generator switch momentarily to RESET, then to ON. Check No. 2 generator voltage.

18. No. 1 and No. 2 generator-out lights - OUT.

19. DC voltmeter selector switch - STRAIN-GAGE.

20. DC voltmeter switch - BUS.

   Check voltage on dc voltmeter.

SAS IN-FLIGHT CHECK.

Before launch, perform the following functional check of the SAS system.

1. "Yar" function switch - STDBY.

2. Gain selector knobs - Set.

   Set the gain selectors at the following: yaw, 8; pitch, 6; roll, 8.

3. Yaw, pitch, and roll function switches - ENGAGE.

   Check that the yaw, pitch, and roll caution lights are out.

4. SAS test switch - WORK, then OFF.

   Move SAS test switch to WORK and check that the SAS yaw, pitch, and roll caution lights blink continuously; then release switch to OFF.

5. Yaw, pitch, and roll function switches - OFF, then ENGAGE.

   When the switches are moved to OFF, check that the caution lights burn steadily; then move the switches to ENGAGE, and check that caution lights go out.

6. SAS test switch - MON, then OFF.

   Move SAS test switch to MON, and check that the SAS pitch yaw, pitch, and roll caution lights blink continuously; then release switch to OFF.

7. Yaw, pitch, and roll function switches - OFF, then ENGAGE.

   When switches are moved to OFF, check that the caution lights burn steadily; then move the switches to ENGAGE, and check that the caution lights go out.

NOTE

If a SAS malfunction is suspected during flight, the pilot can perform the preceding check at his discretion. The SAS check may be performed on any one function or a combination of pitch, roll, and yaw functions.

COUNTDOWN.

1. Liquid oxygen transfer switch OFF (performed by launch operator) - Check.

2. (Deleted)

3. Instrumentation switches - ON.

4. Ball nose test button - Depress; then release.

   Depressing the ball nose test button electrically simulates a predetermined airplane attitude, and the ball nose should drive to a position that cancels out the signal and causes the angle-of-attack indicator to show about a 5-degree nose-down indication. The sideslip indicator will also read about a 15-degree sideslip to the left. When the button is released, the ball nose should drive to the extreme position and appear as a 40-degree nose-up indication on the attitude indicator and a 30-degree sideslip to the right on the sideslip indicator. These readings should be maintained for 2 to 3 seconds; then the ball nose should resume normal operation, driving rapidly without overshoot to indicate the actual angle of attack and sideslip of the airplane.

NOTE

Since the actual airplane attitude may approximate that of the error signal when the button is pressed, positive indications of proper ball nose operation are full-scale readings obtained on the indicators when the button is released and the subsequent return to normal angle-of-attack and sideslip indications.

2-8 CHANGED 26 APRIL 1963

5. Instrument readings - Check. See figure 2-4 for proper instrument reading before pressurization.

6. Vent, pressurization, and jettison lever - JETTISON.

   The tests will be conducted concurrently on all three systems (liquid oxygen, ammonia, and hydrogen peroxide).

7. Jettison stop switches - JETT.

   Launch operator will visually check for vapor emitting at the jettison port, and notify X-15 Airplane pilot.

8. Vent, pressurization, and jettison lever - PRESSURIZE.

   When the vent pressurization, and jettison lever is moved to PRESSURIZE, ammonia and liquid oxygen will be supplied to the turbopumps. Hydrogen peroxide will be supplied to the pump emergency cutoff valve.

9. Liquid oxygen top-off flapper valve - Check.

   Obtain confirmation from the launch operator that vaporous oxygen is no longer emitting from the liquid oxygen overboard vent of the X-15 Airplane pylon.

   If liquid oxygen continues to emit from the pylon vent, indicating that the flapper valve is stuck open, the launch must be aborted.

10. SAS function switches - ON.

11. Flight controls - Check.

    Move all flight controls through allowable travel; receive verbal acknowledgment from the launch operator that all controls are operating properly. Refer to "Launch Limitations" in Section V for allowable travel.

12. Trim system - Check.

    Move horizontal stabilizer through allowable trim travel; receive verbal acknowledgment from the launch operator that trim is operating the stabilizer properly. Refer to "Launch Limitations" in Section V for allowable trim settings.

13. Launch trim - Set.

    Reset launch trim; receive acknowledgment from launch operator.

14. Chamber and stage 2 igniter pressure gage - Check, (both pointers, 0 psi).

15. Data recorder - ON.

16. Engine precool switch - PRECOOL.

    The engine will continue to prime (at high flow rates) until the actual start stops the prime.

17. Engine prime switch - PRIME.

    Move engine prime switch to PRIME for one second, and check ignition ready light ON. Approximately 30 seconds is required to prime, with the prime valve at high-flow orifice position.

NOTE

- If the engine precool switch is off, the engine prime valve is actuated to the low-flow orifice position.

- The prime can be stopped at any time by placing the engine prime switch at STOP PRIME momentarily, then releasing it. This closes the liquid oxygen and NH_3 tank main propellant valves and the H_2O_2 safety valve. The engine is then automatically purged for 17 seconds.

18. Turbopump idle button - Depress for one second.

19. Telemeter and radar switches - Recheck.

20. Communications - Check.

    Check communication with ground station, carrier pilot, and chase pilot.

21. Instruments - Check.

    After pressurization and before launch, check instruments for proper readings. See figure 2-5.

22. Ready-to-launch switch - ON.

    Verbally check with carrier pilot and launch operator that the ready-to-launch light is on.

23. Igniter idle switch - IGNITER.

    Operation of igniter idle is limited to 30 seconds. When the igniter idle switch is placed to IGNITER, the ignition-ready light goes out for 2 seconds while the engine is purged with helium and the igniter spark plugs are energized. When this phase is completed, the ignition-ready light comes on again. When 7 seconds remain of the normal igniter idle phase, the no-drop caution light will come on. With the no-drop caution light on, the pilot must terminate the igniter idle phase (by moving the engine prime switch to STOP PRIME) or continue on to the launch phase.

COCKPIT INSTRUMENT READING BEFORE PROPELLANT SYSTEM PRESSURIZATION (APU'S OPERATING)

All readings are based on requirements for a full-duration mission.

INSTRUMENT		LOW	NORMAL	HIGH	REMARKS
PROPELLANT SOURCE PRESSURE GAGE		3000	3500 to 3600	3900	
H_2O_2 SOURCE AND PURGE PRESSURE GAGE (BOTH POINTERS)		2500	3500 to 3600	3900	Low limit is minimum pressure required for full-duration mission. High limit is maximum relief valve setting.
APU SOURCE PRESSURE GAGE (BOTH POINTERS)		3000	3500 to 3900	4200	
APU H_2O_2 TANK PRESSURE GAGE (BOTH POINTERS)		500	550 to 610	630	Pressure above high limit indicates faulty regulator.
CABIN HELIUM SOURCE PRESSURE GAGE		3000	3500 to 3600	3900	Low limit is minimum pressure required for 30 minutes operation of air conditioning and pressurization system. High limit is maximum relief valve setting.
HYDRAULIC TEMPERATURE GAGES		$-100°C$	$-75°C$ to $0°C$	$95°C$	Low limit is minimum temperature for APU starting.
HYDRAULIC PRESSURE GAGE (BOTH POINTERS)		2600 (AT MAX FLOW DEMAND)	3200 to 3400 (STATIC)	4500	4500 is pump relief valve setting and is acceptable for 30 seconds after APU start.
			2900 to 3400 (WITH FLOW DEMAND)	3500	3500 is maximum after 30 seconds after APU start.
APU BEARING TEMPERATURE GAGE (BOTH POINTERS)		$25°C$	$80°C$ to $130°C$	$150°C$	
MIXING CHAMBER TEMPERATURE GAGE (BOTH POINTERS)		$-60°C$	$-45°C$ to $-35°C$	$-25°C$	Continuous low temperature operation indicates possible injector malfunction, using excessive liquid nitrogen. Temperature increase above $-25°C$ indicates possible injector malfunction or liquid nitrogen depletion. Liquid nitrogen depletion will best be indicated by rapid decrease of cabin helium source pressure.
AC VOLTMETERS		190	195 to 205	210	Limits shown are required or proper operation of airplane components.
LIQUID OXYGEN BEARING TEMPERATURE GAGE		$-40°C$	$-10°C$ to $30°C$	$30°C$	
H_2O_2 TANK AND ENGINE CONTROL LINE PRESSURE GAGE	"C" POINTER	575	575 to 600	620	High limit indicates excessive regulator leakage.
	"T" POINTER		0		Tank vented.
PROPELLANT TANK PRESSURE GAGE	"A" POINTER	0	10 to 30	68	High limit indicates regulator leakage. Vent valve manually closed.
	"L" POINTER		0 to 5		Tank vented.

Figure 2-4.

COCKPIT INSTRUMENT READINGS AFTER PROPELLANT SYSTEM PRESSURIZATION

(APU'S OPERATING)

NOTE
All readings are based on requirements for a full-duration mission.

INSTRUMENT		LOW	NORMAL	HIGH	REMARKS
PROPELLANT SOURCE PRESSURE GAGE		2500	3300 to 3400		Low limit is minimum pressure required for full-duration operation.
H_2O_2 SOURCE AND PURGE PRESSURE GAGE (BOTH POINTERS)		2400	2700 to 3200		
PROPELLANT TANK PRESSURE GAGE (BOTH POINTERS)		45	45 to 58	68	High limit is high setting of primary relief valve.
PROPELLANT PUMP INLET PRESSURE GAGE	"L" POINTER	45	45 to 58	68	
	"A" POINTER	45	45 to 58	68	Pressure readings applicable only after engine prime sequence is initiated.
APU SOURCE PRESSURE GAGE (BOTH POINTERS)		3000	3500 to 3700		Low limit is minimum pressure required for full-duration mission.
APU H_2O_2 TANK PRESSURE GAGE (BOTH POINTERS)		500	550 to 610	630	Pressure above high limit indicates faulty regulator.
CABIN HELIUM SOURCE PRESSURE GAGE		3000	3300 to 3400		Low limit is minimum pressure required for 30 minutes operation of air conditioning and pressurization system.
HYDRAULIC TEMPERATURE GAGES		−75°C	−75°C to 150°C	230°C	Temperature above high limit will cause seal deterioration.
HYDRAULIC PRESSURE GAGE (BOTH POINTERS)		2600	2900 to 3400	3500	High limit is system relief valve setting.
MIXING CHAMBER TEMPERATURE GAGE (BOTH POINTERS)		−60°C	−45°C to −35°C	−25°C	Continuous low-temperature operation indicates possible injector malfunction, using excessive liquid nitrogen. Temperature above −25°C indicates possible injector malfunction or liquid nitrogen depletion. Liquid nitrogen depletion will best be indicated by rapid decrease of helium source pressure.
AC VOLTMETERS		190	195 to 205	210	Limits shown are required for proper operation of airplane components.
APU BEARING TEMPERATURE GAGE (BOTH POINTERS)		25°C	80°C to 130°C	150°C	
H_2O_2 TANK AND ENGINE CONTROL LINE PRESSURE GAGE (BOTH POINTERS)		550	575 to 600	620	High limit indicates above-normal regulator pressure. Pressures below 575 will result in pressures below the engine specification inlet requirements for full-thrust missions.
LIQUID OXYGEN BEARING TEMPERATURE GAGE		−40°C	−10°C to 30°C	30°C	If bearing temperature is below low limit, launch must be aborted.

Figure 2-5

CAUTION

The igniter idle phase must be terminated immediately if the idle-end caution light comes on, as damage to the engine chamber will occur because of insufficient cooling.

24. Chamber and stage 2 igniter pressure gage - Check (short hand, 150 psi in approximately 5 seconds).

25. Ready-to-launch - Countdown by carrier pilot.

BALLISTIC CONTROL AND REACTION AUGMENTATION SYSTEM OPERATION.

Since most missions will involve flight at altitudes where ballistic control system operation will be required to maintain airplane attitude, the ballistic control system should be turned on before launch, in order to be available for use when required. The reaction augmentation system should be turned on as soon as possible after engine burnout. To turn on the ballistic control and reaction augmentation systems, proceed as follows:

1. No. 1 and No. 2 ballistic control switches - ON.
2. Automatic cutoff switch - ON.
3. Reaction augmentation function switches - ENGAGE.
4. RAS-out light - Check out.

LAUNCH.

During launch, the following X-15 Airplane control surface deflections are recommended:

Stabilizer	
Symmetrical deflection	0° to -2° (stabilizer leading edge down)
Differential deflection	0°
Vertical stabilizer	0°
Speed brakes	0° (in)
Flaps	0° (up)

After release from the carrier airplane, proceed as follows:

1. Throttle - Inboard to 50% (40%).

    Throttle must be moved to 50% (40%) by the time the idle-end caution light comes on.

2. Chamber and stage 2 igniter pressure gage - Check.
3. Propellant manifold pressure gage - Check.
4. Propellant tank pressure gage - Check.
5. H_2O_2 tank and engine control line pressure gage - Check.

FLIGHT CHARACTERISTICS.

Refer to Section VI for information regarding flight characteristics.

ENGINE OPERATION.

NORMAL INDICATIONS DURING START.

When the thrust chamber is fired, the following indications will be evident in about the sequence given. These indications occur in rapid sequence (about 5 seconds).

1. Turbine whine.
2. Turbine exhaust steam will be seen at the exhaust overboard discharge line.
3. Fuel and liquid oxygen manifold pressure will rise to rated values and will be stabilized by the pump governor.
4. Igniters will be operating.
5. Pressure of the chamber will rise to a point where chamber pressure will be shown on the indicator gage.

    The liquid oxygen and fuel will automatically stop bleeding overboard.

6. Chamber pressure will reach rated value.
7. Thrust chamber will emit a great deal of noise.

ENGINE THRUST CONTROL.

Engine thrust is controlled by movement of the throttle between 50% (40%) and 100% thrust. Engine response to throttle movement is very rapid, 40% to 100% in approximately 1.5 seconds.

NORMAL OPERATING CONDITIONS.

See figure 2-6 for proper instrument readings for all systems during powered flight.

ENGINE BURNOUT.

Propellant exhaustion will result in the following:

1. Fuel or oxygen manifold pressure drops, with consequent shutdown by low manifold pressure safety circuit.
2. Fuel manifold pressure drops below oxygen pressure.
3. Engine runs rough or jet flame is unsteady.
4. Pump cavitates with consequent overspeed cutoff.

COCKPIT INSTRUMENT READINGS IN FLIGHT
(ENGINE OPERATING)

NOTE
- All readings are based on requirements for a full-duration mission.
- Figures in parentheses are for 40% thrust configuration.

INSTRUMENT		LOW	NORMAL	HIGH	REMARKS
H₂O₂ SOURCE AND PURGE PRESSURE GAGE (BOTH POINTERS)		800	3000 (GRADUALLY DECREASING)		Pressure below low limit may result in erratic engine operation.
PROPELLANT TANK PRESSURE GAGE (BOTH POINTERS)		40	45 to 53	68	Upper limit is high setting of pressure relief valve.
PROPELLANT PUMP INLET PRESSURE GAGE	"L" POINTER	35	40 to 70		
	"A" POINTER	35	40 to 55		
APU H₂O₂ TANK PRESSURE GAGE (BOTH POINTERS)		500	550 to 610	630	
CABIN HELIUM SOURCE PRESSURE GAGE			1000 to 3400		Liquid nitrogen nearing depletion when pressure reaches approximately 1000.
HYDRAULIC TEMPERATURE GAGES		−55°C	0°C to 150°C	230°C	Temperature above high limit can cause seal deterioration.
HYDRAULIC PRESSURE GAGE (BOTH POINTERS)		2600	2900 to 3400	3500	Low limit is minimum pressure required for surface deflection at maximum load. High limit is system relief valve setting.
APU BEARING TEMPERATURE GAGE (BOTH POINTERS)		25°C	80°C to 130°C	200°C	Temperature above high limit may cause damage to APU.
MIXING CHAMBER TEMPERATURE GAGE (BOTH POINTERS)		−60°C	−45°C to −35°C	−25°C	Continuous low-temperature operation indicates possible injector malfunction, using excessive liquid nitrogen. Temperature above −25°C indicates possible injector malfunction or liquid nitrogen depletion. Liquid nitrogen depletion will best be indicated by rapid decrease of cabin helium source pressure.
AC VOLTMETERS		190	195 to 205	210	Limits shown are required for proper operation of airplane components.
H₂O₂ TANK AND ENGINE CONTROL LINE PRESSURE GAGE	"C" POINTER	550	575 to 600	620	Pressure below 450 may result in malfunction shutdown. Pressure exceeding high limit indicates regulator malfunction.
	"T" POINTER	550	565 to 600	620	Pressure below 565 is below minimum engine specification H₂O₂ inlet requirements for full-thrust missions.
PROPELLANT MANIFOLD PRESSURE GAGE	"L" POINTER	380 (310)	440 to 1050 (350 to 1050)	1200	Pressure above high limit indicates injector blockage. Operation below low limit will result in malfunction shutdown.
	"A" POINTER	435 (341)	495 to 1150 (391 to 1150)	1300	
CHAMBER AND STAGE 2 IGNITER PRESSURE GAGE	LONG POINTER	310 (240)	345 to 600 (265 to 600)	630	Operation outside limits shown may result in engine damage.
	SHORT POINTER	315 (245)	350 to 630 (275 to 630)	660	During operation, second-stage igniter pressure should be above main chamber pressure.

X-15-1-51-13A

Figure 2-6

Section II

SHUTDOWN PROCEDURE.

To shut down engine, proceed as follows:

1. Throttle - OFF.

   Retard throttle to 50% (40%); then move outboard to OFF.

   #### NOTE

   The ignition-ready light will go out for approximately 5 seconds. During this time, the engine is purged with helium. After purging, the ignition-ready light comes on. The engine is now on stand-by prime condition. Restart can be made by moving throttle to 50% (40%) position.

2. Igniter idle switch - OFF.

3. Engine prime switch - STOP PRIME momentarily, then release.

4. Engine master switch - OFF.

5. Vent, pressurization, and jettison lever - VENT.

ABORTED LAUNCH.

If for any reason the decision is made to abort the launch after the countdown has started, proceed as follows:

1. Prime switches or switch - OFF or STOP PRIME.

2. Engine master switch - OFF.

3. No. 1 and No. 2 generator switches - OFF.

   Check that both No. 1 and No. 2 generator-off lights come on.

4. APU switches - OFF.

5. Oxygen selector - B-52.

   Reselect B-52 oxygen supply, and check.

6. Ventral arming switch - DE-ARM.

7. Carrier pilot - Notify.

   Notify carrier pilot and launch operator when launch abort procedure is completed.

CAPTIVE JETTISON.

To jettison fuel from the X-15 Airplane after an aborted launch, proceed as follows:

1. Source pressure - Check.

2. Jettison stop switches - Recheck JETT.

3. Vent, pressurization, and jettison lever - JETTISON.

   Check jettison by pressure bleed on the source pressure gages. Stop jettisoning when propellant source pressure reaches 600 psi.

4. Jettisoning - Check.

   Have chase pilot verify that fuel is jettizoning.

5. Vent, pressurization, and jettison lever - VENT.

   After propellant has been jettisoned, move control to VENT.

DESCENT.

Normally, the descent will be prebriefed, however, it will be constantly monitored and controlled from the ground.

CAUTION

Because of the high rate of descent and the reduced stability at low Mach numbers, the speed brakes are not to be used at full deflection below Mach 1.5.

BEFORE LANDING.

During descent, just before entering the landing pattern, check all controls and instruments for landing.

NOTE

Before landing, preferably on the downwind leg of the landing pattern, but in no case above 17,000 feet above sea level, move vent, pressurization, and jettison lever to PRESSURIZE, to prevent sand and dust from entering the airplane propellant system during landing. The altitude limitation is necessary to preclude structural deformation of the airplane propellant tanks due to a pressure differential which would tend to collapse the tanks.

LANDING.

See figure 2-7 for the recommended landing pattern and procedures. See figure 2-8 for the recommended low speed, low altitude landing pattern.

CAUTION

- To provide ground clearance, the lower ventral must be jettisoned before landing. The ventral should be jettisoned a minimum of 1500 feet above the terrain.

- Directional control is reduced when the lower ventral is jettisoned; however, adequate directional control is available for landing.

AFTER LANDING.

After landing, as soon as the airplane stops, proceed as follows:

1. Canopy - Open.

CHANGED 26 APRIL 1963

2. Face plate - Remove.

3. Ram-air door - Close.

4. Wing flap switch - UP.

5. SAS function switches - STDBY.

6. Ventral arming switch - DE-ARM.

7. External power - ON.

8. APU switches - OFF.

Bleed off hydraulic system pressure. This will help prevent accumulator failure due to slow bleed-down of hydraulic system pressure, especially at elevated temperatures.

9. Speed brake levers - Full aft position.

Before operating the speed brakes, be sure the fuselage rear section around the speed brakes is clear, because the brakes operate rapidly and forcefully and could injure any personnel near the brakes.

10. Center control stick - Full forward.

11. Rudder pedals - Actuate.

Deplete hydraulic pressure by actuating rudder pedals.

BEFORE LEAVING AIRPLANE.

Before leaving the airplane, complete the required airplane forms and then verify the following cockpit control positions:

LEFT CONSOLE.

1. Radio controls - OFF.

2. Intercom control - B-52.

3. Vent suit heater switch - OFF.

3A. Alternate SAS switch - OFF.

4. Wing flap switch - UP.

5. (Deleted)

6. Trim control switch - NORMAL.

7. Ready-to-launch switch - OFF.

8. Jettison trim switch - OFF.

9. Speed brake handles - Full forward.

10. Vent, pressurization, and jettison lever - VENT.

11. Throttle - OFF.

12. Jettison stop switches - STOP.

13. Auxiliary launch switch - OFF.

14. Landing gear handle - IN.

15. Face mask heater alternate switch (on ejection seat) - OFF.

INSTRUMENT PANEL.

1. Engine master switch - OFF.

2. Emergency battery switch - OFF.

3. Generator switches - OFF.

4. APU switches - OFF.

4A. Hydrogen peroxide transfer switch - OFF.

5. Ballistic control switches - OFF.

6. APU water cooling switches - OFF.

CENTER PEDESTAL.

1. SAS function switches - STDBY.

2. Research instrumentation - OFF.

3. Ram-air lever - CLOSED.

4. Cockpit ram-air knob - CLOSED (in).

5. Radar beacon switch - OFF.

6. Stable platform instrument switch - ON.

RIGHT CONSOLE.

1. Canopy emergency release handle - IN.

2. Stable platform switch - OFF.

3. Nose ballistic rocket heater switch - OFF.

4. Cockpit lighting switch - OFF.

5. Indicator, caution, and warning light switch - NORMAL.

6. Windshield heater switches (two) - OFF.

7. Pressure cooling lever - OFF.

8. Blower switches - OFF.

9. All circuit breakers - OFF.

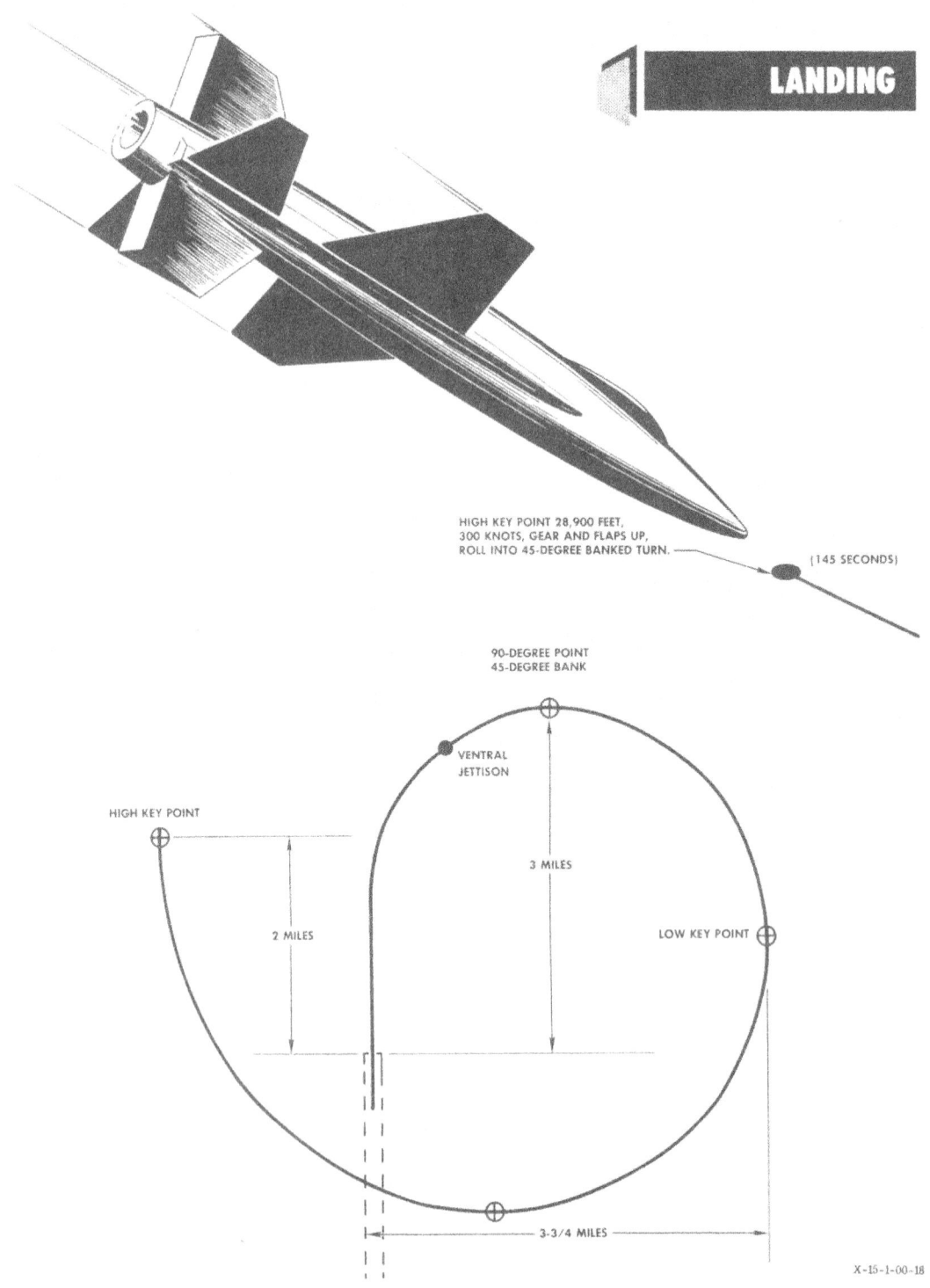

Figure 2-7

PATTERN (TYPICAL)

NOTE

- Before landing, preferably on the downwind leg of the landing pattern, but in no case above 17,000 feet above sea level, move vent, pressurization, and jettison control lever to **PRESSURIZE,** to prevent sand and dust from entering the airplane propellant system during landing. The altitude limitation is necessary to preclude structural deformation of the airplane propellant tanks due to a pressure differential which would tend to collapse the tanks.
- To ensure safe recovery of the ventral, the ventral should be jettisoned 1500 feet above the terrain.
- If the high key point for landing cannot be reached, reduce airspeed to 240 knots IAS until subsequent key point can be reached. Then dive to increase airspeed to 300 knots IAS.

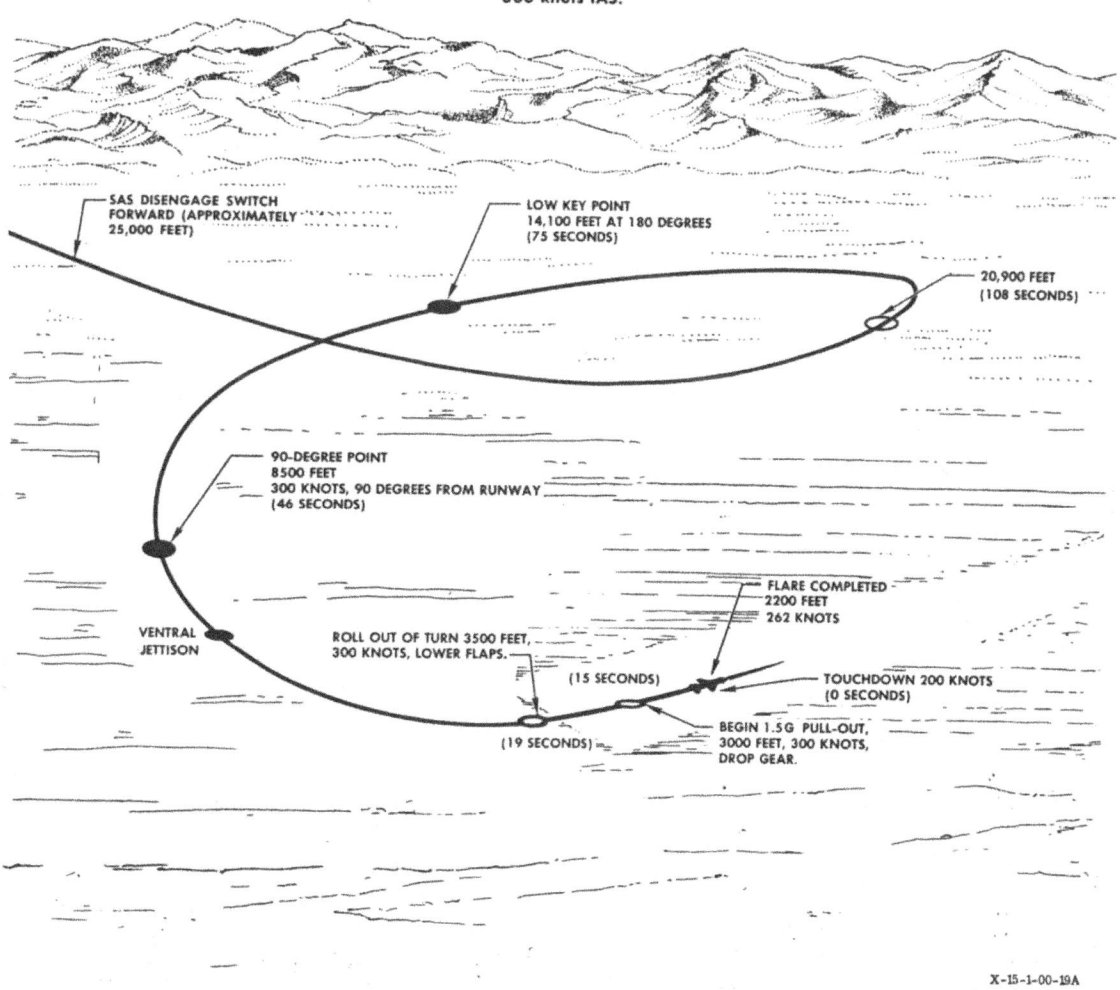

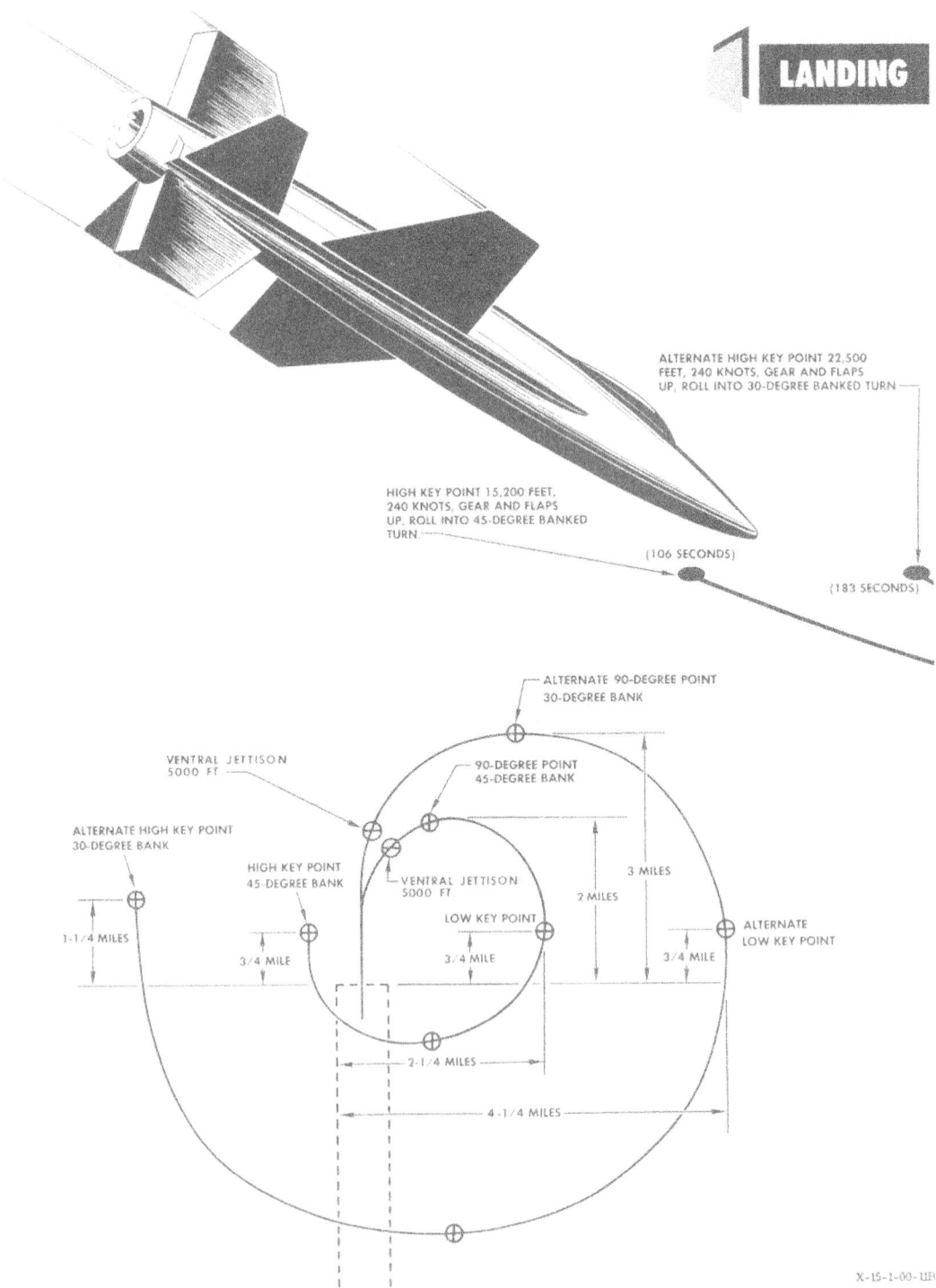

Figure 2-8

PATTERN (LOW SPEED—LOW ALTITUDE)

NOTE

- Between 25,000 feet and the high key point, move the SAS DISENGAGE switch forward.
- Before landing, preferably on the downwind leg of the landing pattern, but in no case above 17,000 feet above sea level, move vent, pressurization, and jettison control lever to **PRESSURIZE**, to prevent sand and dust from entering the airplane propellant system during landing. The altitude limitation is necessary to preclude structural deformation of the airplane propellant tanks due to a pressure differential which would tend to collapse the tanks.
- To ensure safe recovery of the ventral, the ventral should be jettisoned 1500 feet above the terrain.

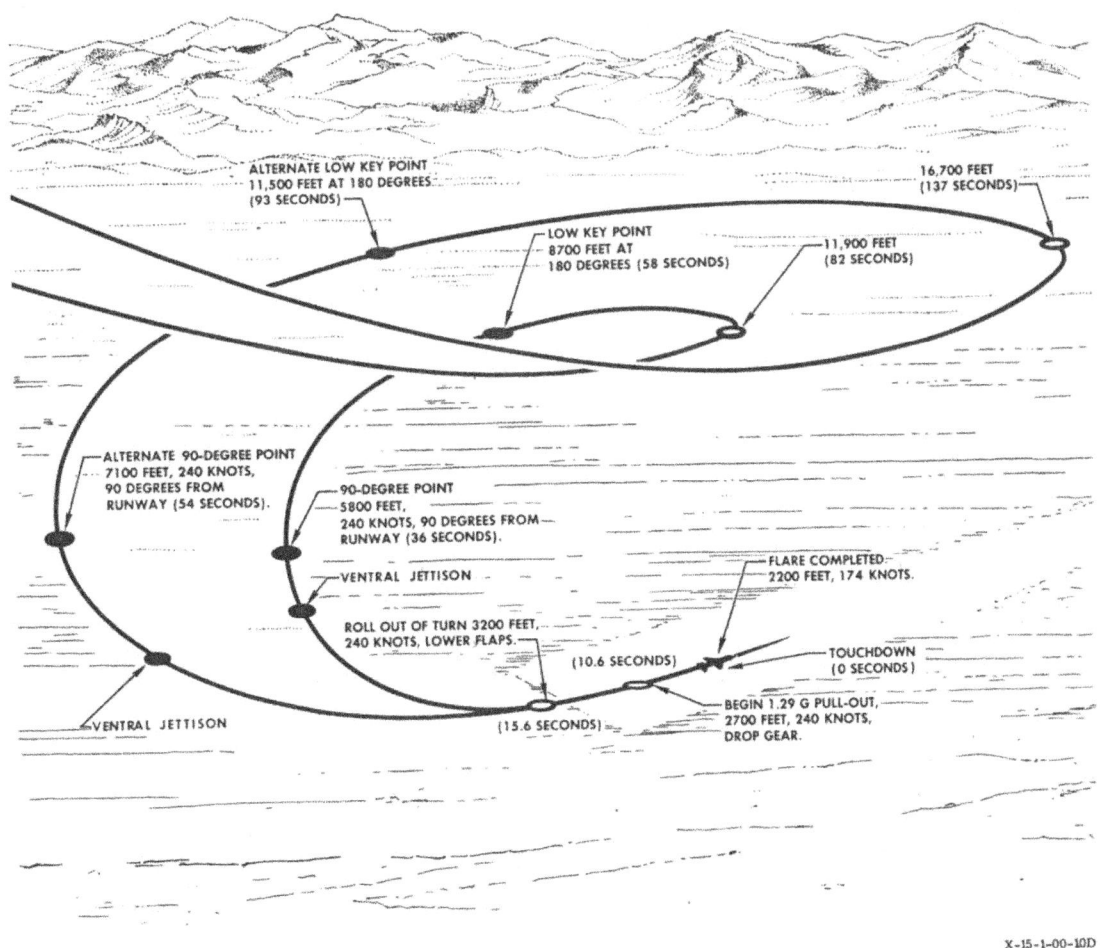

THIS PAGE INTENTIONALLY LEFT BLANK.

T.O. 1X-15-1

Emergency Procedures

SECTION III

TABLE OF CONTENTS	PAGE		PAGE
Engine Failure	3-1	Emergency Entrance	3-5
Captive Operation Emergencies	3-2	Fuel System Failure	3-5
Emergency Launch	3-2	Turbopump Failure During Flight	3-5
Maximum Glide	3-3	Propellant Jettisoning	3-8
Forced Landing	3-3	APU System Failure	3-8
Fire or Explosion	3-3	Electrical Power Supply System Failure	3-8
Elimination of Smoke or Fumes	*	Hydraulic System Failure	3-9
Ejection	3-5	Flight Control Hydraulic System Failure	3-9
Ventral Jettison System Failure	3-5	Ballistic Control System Failure	3-14

ENGINE FAILURE.

Failure of rocket engines is, as a rule, the result of thrust chamber burnout. Specific information on this type engine failure is given in "Thrust Chamber Burnout" in this section. If engine failure is due to malfunction of the fuel control system or improper operating technique, a start can usually be made to restore engine operation, provided time and altitude permit. However, if the failure is an obvious failure within the engine, a start should not be attempted.

MALFUNCTION SHUTDOWNS.

Automatic malfunction shutdowns can occur during the engine start phase or during actual engine operation. The design of the malfunction shutdown circuits is such that restart attempts can safely be made after a malfunction shutdown. This is because a restart will not be successful if the malfunction which caused the original shutdown has not been corrected. Malfunction shutdowns which can occur during the start phase are those due to main or first stage propellant valve malfunction, stage 2 ignition malfunction, or engine turbopump overspeed. Malfunction shutdowns which can occur during engine operation are those caused by engine turbopump overspeed or excessive engine vibration. If a malfunction shutdown occurs, attempt a restart. Refer to "Engine Restart" in this section.

Causes of Pump Cavitation.

Failure to sufficiently prime the liquid oxygen pump discharge line to the engine, excessive heat absorption by the liquid oxygen together with excessive vapor pressure (normal liquid oxygen tank vapor pressure after tank has been filled is 20 to 24 psia), or insufficient liquid oxygen tank pressure can cause pump cavitation. When liquid oxygen pump cavitation occurs, the overspeeding turbopump is stopped by the overspeed protection system. When an engine is shut down by the turbopump overspeed protection system, the turbopump overspeed caution light is illuminated.

Indications of Pump Cavitation.

Cavitation is accompanied by the following indications:

- Increase of pitch in whine of turbine.
- No ignition.
- No flame from chamber.
- No noise from chamber.
- No thrust chamber pressure.
- Very high momentary fuel manifold pressure.
- Very low liquid oxygen manifold pressure.
- Turbopump stops and overspeed indicator light, in cockpit, comes on.

*At the time of publication of this flight manual, information for various specific procedures was not available. These will be supplied as soon as available.

Section III   T.O. 1X-15-1

ENGINE RESTART.

If the engine has failed or has been shut down automatically and it is determined that a restart is feasible, proceed as follows:

1. Throttle - OFF.

2. Engine reset button - Push (one second).

3. Engine prime switch - PRIME.

   Move engine prime switch to PRIME for one second and check ignition-ready light ON. Approximately 30 seconds is required to prime, with prime valve at high-flow orifice position.

4. Throttle - As desired.

THRUST CHAMBER BURNOUT.

Thrust chamber burnout can be determined by various engine indications which can be verified by the pilot or the chase pilot. These indications are as follows:

a. Fuel manifold pressure drops below oxygen manifold pressure, accompanying cylinder "groaning."

b. Engine emits loud scream or howl.

c. Yellow streak of flame or yellow bushy flame is evident.

d. Chamber pressure drops while manifold pressures are within specified range.

CAPTIVE OPERATION EMERGENCIES.

BEFORE TAKE-OFF.

In case a "FIRE" warning, "APU COMPT HOT," or "H_2O_2 HOT" light comes on before take-off, notify the carrier pilot that an X-15 emergency is in progress (refer to "Fire or Explosion" in this section), and prepare to abandon the airplane.

IN-FLIGHT.

In case of an emergency during captive flight, the carrier pilot must verify the emergency and ascertain the danger to the crew. It will be the carrier pilot's responsibility to initiate one or more of the following emergency procedures:

   a. X-15 emergency launch. (Refer to "Emergency Launch" in this section.)

   b. X-15 pilot ejection. (Refer to "Ejection" in this section.)

   c. Abort mission or launch. (Refer to "Aborted Launch" in Section II.)

The five basic rules covering selection of the proper procedure are as follows:

   a. X-15 hazardous to B-52 - Launch X-15. (This may be preceded by X-15 pilot ejection if the X-15 is hazardous to the X-15 pilot.)

   b. B-52 hazardous to X-15 below critical altitude - Eject X-15 pilot and B-52 crew.

   c. B-52 hazardous to X-15 above critical altitude - Launch X-15 and eject B-52 crew.

   d. X-15 emergency not immediately hazardous - Return to base for landing.

   e. B-52 emergency not immediately hazardous - Return to base for landing.

EMERGENCY LAUNCH.

In case of an emergency aboard the B-52 which endangers the X-15 or in case of an emergency in the X-15 which would endanger the B-52, it is possible to emergency-launch the X-15. If time and altitude permit a launching of this type, proceed as follows:

1. Oxygen system - Switch from carrier supply to X-15 Airplane supply.

2. Ventral arming switch - ARM.

3. Emergency battery - ON.

   If APU's are not operating at the time of emergency launch, the emergency battery must be on to enable APU start.

4. No. 1 and No. 2 APU switches - ON.

   Monitor hydraulic gages for both systems as APU's come up to speed.

5. No. 1 and No. 2 generator switches - To RESET momentarily, then to ON.

6. Ram-air lever - Recheck CLOSED.

7. Cockpit helium switch - ON.

8. Pressure-cooling lever - ON.

9. Blower switches - ON.

10. Launch light - On.

11. SAS function switches - ENGAGE.

12. Ventral jettison button - Push.

13. Instrumentation master switch - ON.

14. Data switch - ON.

15. Vent, pressurization, and jettison lever - JETTISON.

MAXIMUM GLIDE.

Glide distance may be varied with Mach to achieve the desired distance. During gliding flight, high rates of descent will be experienced; speed brakes will increase the rate of descent considerably and should only be used at speeds above Mach 1.5. Speeds above Mach 1.5 will usually occur at high altitude and will allow sufficient time to reduce the rate of descent before entering the landing phase. See figure 3-1 for maximum glide for given initial Mach numbers.

FORCED LANDING.

In case of either inability to retract the speed brakes or inadvertent extension of the landing gear before entering the landing pattern, certain precautions must be observed. (See figure 3-2.)

FIRE OR EXPLOSION.

In case of a fire or explosion, the procedures given in the following paragraphs should be accomplished. However, an important factor in determining the course of action to be taken depends on the effect the fire or explosion has on the flight control systems. Since a failure of the aerodynamic flight control system could occur as a result of the fire or explosion, a careful check of the aerodynamic system should be made to determine whether a safe landing can be made.

ENGINE FIRE.

If a fire-warning light comes on or if there are other indications of fire, proceed as follows:

1. Nitrogen or helium release selector switch - Check AUTO.

   With the nitrogen or helium release selector switch at AUTO when the fire-warning system is activated, the engine is automatically shut down and the engine compartment purged. If automatic shutdown is successful, proceed to step 3. However, if the engine fails to shut down, proceed to step 2.

2. Shut down engine.

   a. Engine master switch - OFF.

   Moving the engine master switch to OFF purges the engine with helium for approximately 17 seconds.

   b. Tank shutoff and N_2 bleed switch - OFF.

   c. Nitrogen or helium release selector switch - ON.

3. Throttle - OFF.

4. Igniter idle switch - OFF.

5. If fire cannot be confirmed, land as soon as possible.

   If fire cannot be confirmed or if it goes out, establish controllability and jettison propellants en route to the nearest available base where a landing can be made. Obtain assistance from chase pilot if possible. Reconfirm controllability before descent below safe ejection altitude.

6. If fire is confirmed or if aerodynamic control is lost - Eject.

ELECTRICAL FIRE.

In case of an electrical fire, attempt to isolate fire by moving circuit breakers off one at a time. Allow sufficient time for the indication of fire to cease. When source of fire is found, leave the affected circuit de-energized, and restore power to remaining circuits.

H_2O_2 COMPARTMENT OVERHEAT.

If the "H_2O_2 COMP HOT" light comes on, immediately notify the carrier pilot that an X-15 emergency is in progress. Then prepare for an emergency launch (refer to "Emergency Launch" in this section), in case an emergency launch becomes necessary.

H_2O_2 OVERHEAT (APU).

If the APU "H_2O_2 HOT" light comes on, immediately notify the carrier pilot that an X-15 emergency is in progress. Determine which APU system is overheated and proceed as follows:

1. APU switch - OFF.

   The APU system that is overheated must be shut down.

2. APU switch - JETT.

   Maximum time for jettison is approximately 27 seconds. Launch operator and chase pilot will monitor jettison pattern.

3. Prepare for emergency launch.

   (Refer to "Emergency Launch" in this section.)

H_2O_2 LOW (APU).

If an APU hydrogen peroxide low caution light comes on before launch, notify the carrier pilot that an X-15 emergency is in progress. Shut down affected APU and prepare for an aborted launch. If the light comes on after launch, monitor affected APU system until engine shut down is accomplished; then position hydrogen peroxide transfer switch to ON.

CAUTION

To prevent loss of operative APU, do not jettison engine H_2O_2 following shutdown if the "H_2O_2 TRANS" switch is at ON.

APU COMPARTMENT OVERHEAT.

If the "APU COMP HOT" light comes on, immediately notify the carrier pilot that an X-15 emergency is in progress. Determine which APU system is overheated and proceed as follows:

1. APU switch - OFF.

   The APU system that is overheated must be shut down.

MAXIMUM GLIDE DISTANCE AND TIME
(TO 20,000 FEET)

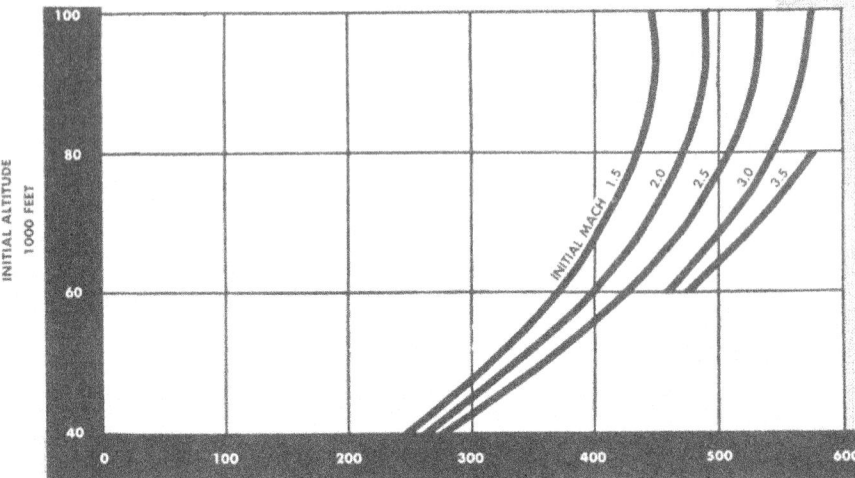

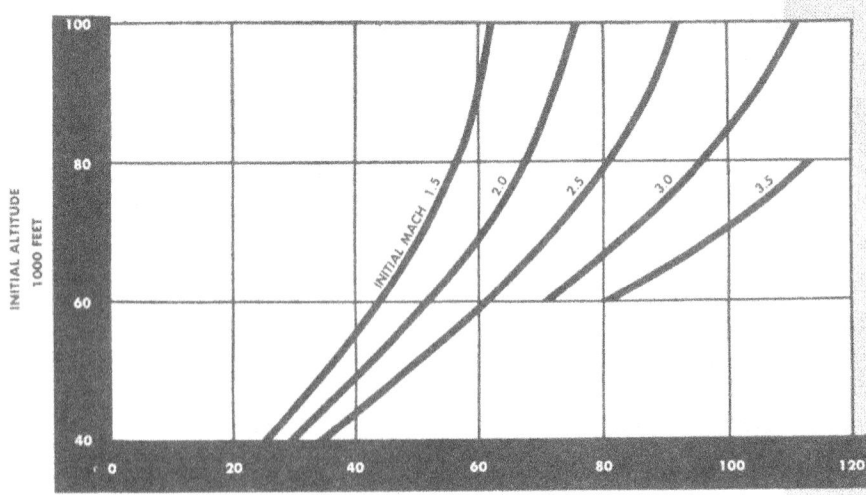

EXAMPLE: From 80,000 feet at Mach 2, 465 seconds will be required to descend to 20,000 feet and 68 nautical miles will be traveled.

Figure 3-1

2. APU switch - JETT.

   Maximum time for jettison is approximately 27 seconds. Launch operator and chase pilot will monitor jettison pattern.

3. Prepare for emergency launch.

   (Refer to "Emergency Launch" in this section.)

EJECTION.

Escape from the airplane in flight must be made by means of the ejection seat. Ejection from 200 knots at zero altitude to any speed and altitude possible by the carrier airplane can be made while the X-15 is attached to the carrier airplane. After launch, ejection is possible up to Mach 4 and any altitude up to 120,000 feet. The basic seat ejection procedure is shown in figure 3-5.

VENTRAL JETTISON SYSTEM FAILURE.

The ventral is jettisoned when the ventral jettison button is pushed. If the ventral fails to jettison, as verified by the chase pilot, pulling the landing gear lowering lanyard should jettison the ventral. Receive verbal confirmation from the chase pilot that the ventral jettisoned.

EMERGENCY ENTRANCE.

Emergency entrance into the cockpit is made by manually opening the canopy, or if necessary, jettisoning it. In either case, the ejection seat must first be disarmed if the ejection handles are raised.

WARNING

- Figure 3-4 shows two procedures for emergency entrance. The procedure to be used depends on the position of the ejection seat handgrips.

- Ammonia, either liquid or gas, is both toxic and flammable and is considered extremely dangerous. It is capable of causing suffocation and severe irritation to the tissue of the eyes, nose, throat, lungs, and skin. In addition to its caustic action, liquid ammonia can also cause severe injury by freezing. Persons exposed to ammonia should be thoroughly familiar with its hazards.

NOTE

Canopy removal by explosive remover is not considered a fire hazard when fuel is depleted or completely jettisoned before landing.

FUEL SYSTEM FAILURE.

AIRPLANE FUEL SYSTEM FAILURE.

If propellant system pressurization is lost, a caution light ("LOX" or "NH3") will come on. System pressure can be restored for continued low thrust engine operation or propellant jettisoning by moving the propellant emergency pressurization switch toward the illuminated light.

NOTE

The emergency pressurization helium supply will provide sufficient emergency pressure for one tank only.

ENGINE FUEL SYSTEM FAILURE.

Failure of the engine fuel system would be the result of an engine turbopump failure. (Refer to "Turbopump Failure During Flight" in this section.)

FUEL LINE PRESSURE LOW.

In the event of a low fuel line pressure caution light during flight (above 40,000 feet), proceed as follows:

1. Throttle - Reduce.

   Reduce engine thrust setting as necessary to maintain fuel inlet pressure at 30 psi. If pressure returns to normal, full throttle operation can be resumed.

2. Continue operation, or shut down.

   Operation can be continued at minimum thrust setting if fuel inlet pressure remains above 20 psi. If fuel inlet pressure cannot be maintained above 20 psi, even at minimum thrust, shut down engine immediately, provided a safe landing can be made.

WARNING

An engine restart is not to be attempted unless pilot or airplane safety necessitates resumption of power.

TURBOPUMP FAILURE DURING FLIGHT.

A turbopump failure will in most cases be caused by pump cavitation, or incorrect governor supply pressure. The possibility of a mechanical malfunction of the relatively simple turbopump assembly is not considered likely. Turbopump failure or shutoff, caused by the overspeed caution light coming on. If the overspeed caution light comes on, the engine may be restarted in the following manner:

1. Throttle - OFF.

2. Engine reset button - Depress (momentarily).

3. Engine precool switch - PRECOOL.

4. Engine prime switch - Check PRIME.

5. Throttle - 50% (40%).

   Move throttle inboard to 50% (40%); then advance as required.

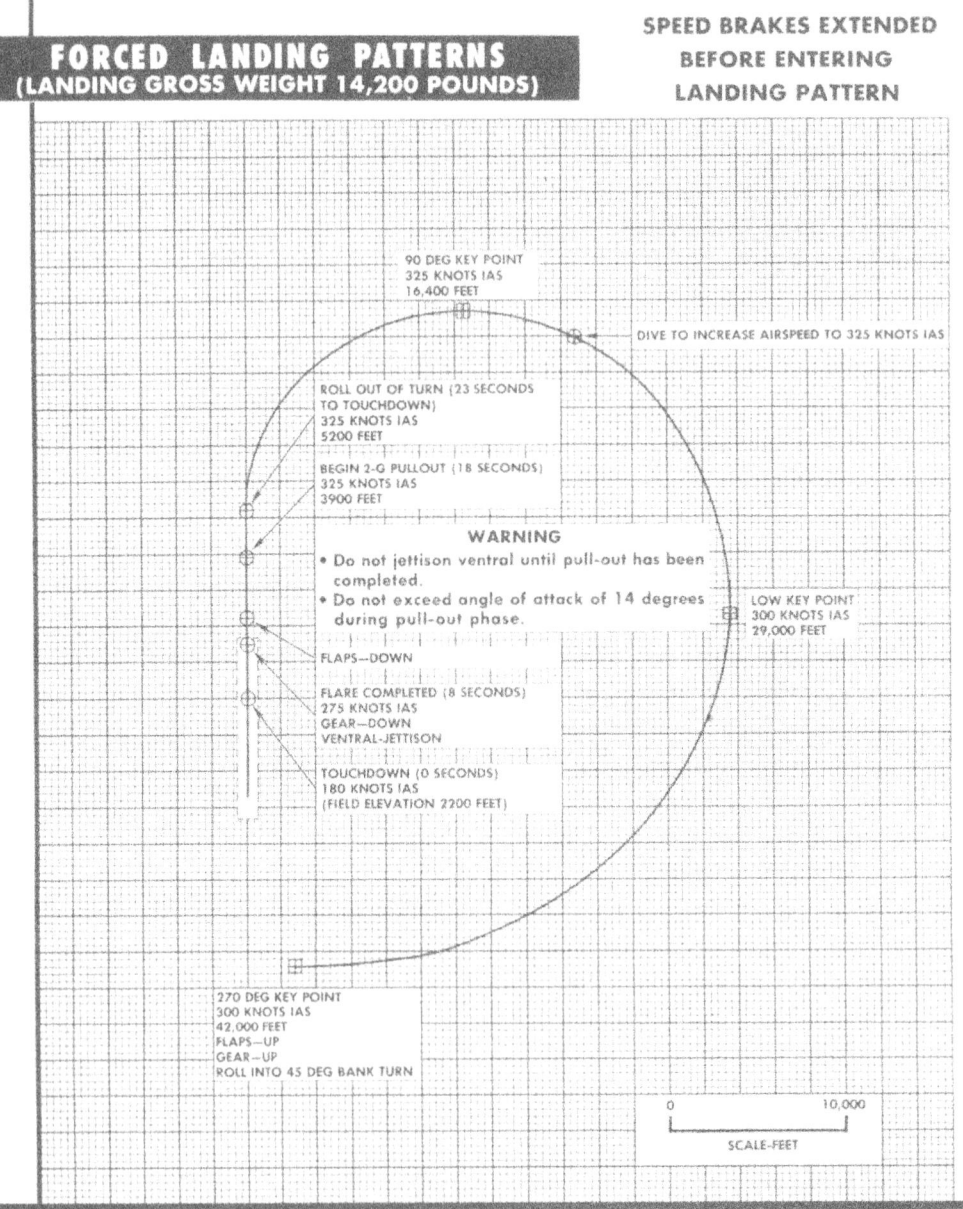

Figure 3-2

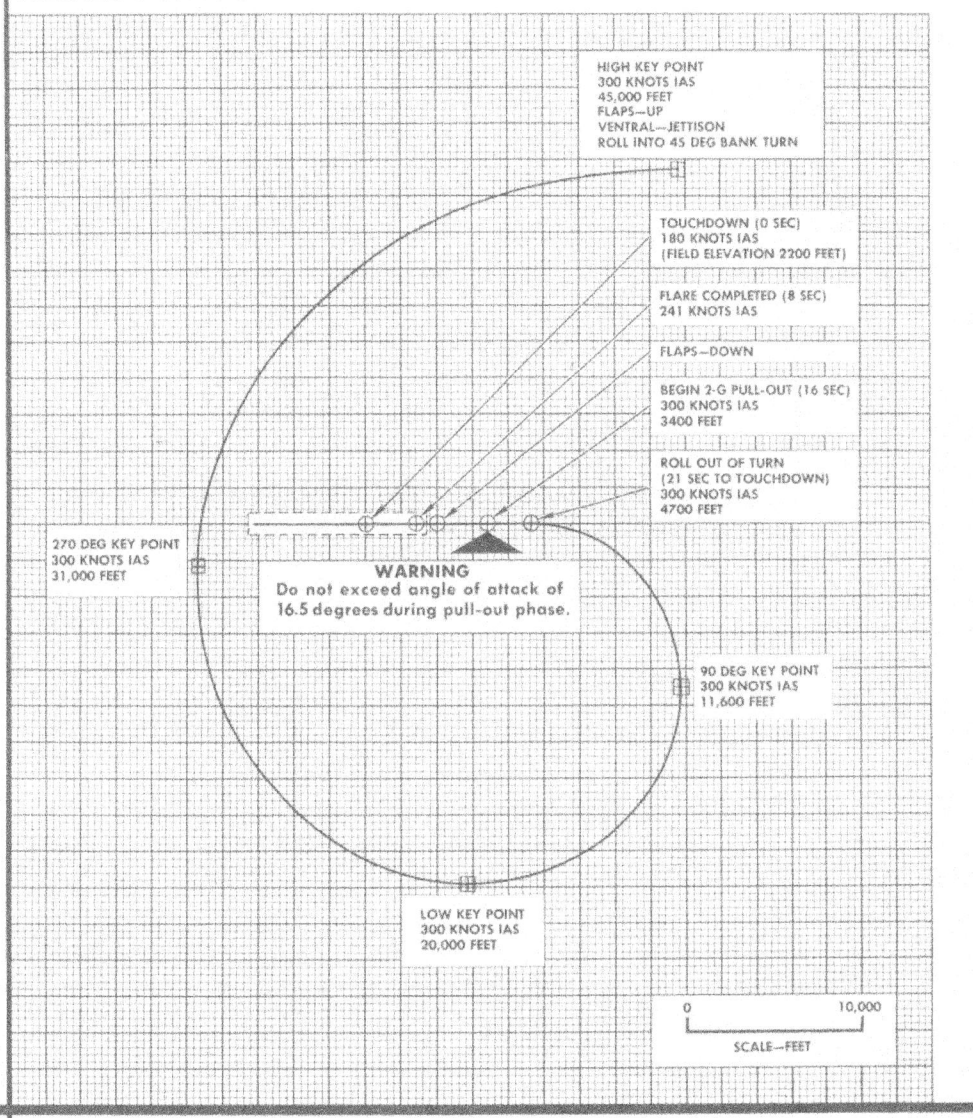

FUEL JETTISON

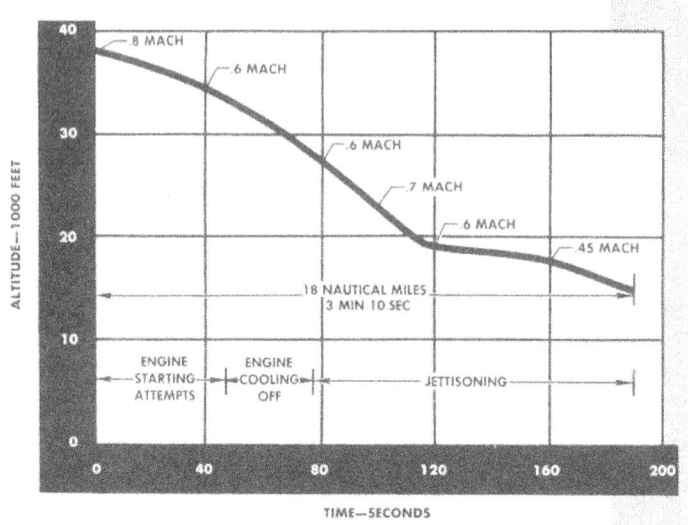

Figure 3-3

PROPELLANT JETTISONING.

In an emergency where the engine propellants must be jettisoned, recheck jettison test switches at JETT; then move the vent, pressurization, and jettison lever to JETTISON, which allows the propellants to flow overboard. Approximate time required to jettison a full propellant load is 120 seconds. (See figure 3-3.)

APU SYSTEM FAILURE.

The dual APU's provided allow normal operation as long as one APU is operating. However, certain equipment not required for safe flight will be lost.

NOTE

The SAS emergency hydraulic system automatically provides hydraulic power for the SAS pitch and roll servos in the event of a No. 2 hydraulic system failure.

LOSS OF	RESULTS IN COMPLETE LOSS OF	NECESSITATES USE OF
APU No. 1	Ball nose SAS yaw	Mission abort may be required.
APU No. 1 & 2	All hydraulic power All electrical power	Emergency battery to attempt to restart APU. If restart fails, ejection is mandatory.

ELECTRICAL POWER SUPPLY SYSTEM FAILURE.

COMPLETE ELECTRICAL FAILURE.

If complete electrical failure occurs because of loss of both generators, the APU's will continue to operate. If the APU's are not operating, the emergency battery must be energized to start the APU's. With both generators failed, the airplane can be landed. If complete electrical failure occurs because of failure of both APU's, the emergency battery switch must be turned ON, so that the APU's can be restarted. If the APU's cannot be restarted, ejection is mandatory.

GENERATOR FAILURE.

If either APU or generator fails and cannot be reset, certain missions may be flown, dependent on ground control. If a failure of this type should occur, proceed as follows:

1. No. 1 or No. 2 blower switch - OFF.

2. Vent suit heater switch - OFF.

3. Pressure suit ventilation knob - OFF.

   As soon as the unique research portion of the flight is over, continue with the following steps.

4. Instrumentation master switch - OFF.

5. Radar beacon circuit breaker - Pull.

6. Telemeter master and commutator switches - OFF.

7. No. 1 and No. 2 blower switches - BLOWER & LN_2.

NOTE

Operation of one equipment compartment blower at a time will substantially reduce the electrical load; however, operation of only one blower will appreciably shorten its life.

GENERATOR RESET PROCEDURE (APU RUNNING).

When a generator irregularity occurs, as shown by a generator failure light coming on, try to bring the generator back as follows:

1. Generator switch - RESET momentarily; then ON.

   Hold generator switch at RESET momentarily; then return switch to ON. Generator-out light should go out and remain out if failure was temporary.

2. Generator-out light - Check.

   If light is still on when generator switch is returned to ON, repeat reset procedure several times. If light remains on after several reset attempts, check voltmeter to determine generator output. Then proceed to step 3.

3. Voltmeter - Check.

   a. If voltmeter indication is normal, continue flight (leave generator switch ON).

   b. If voltmeter indicates no charge, continue flight. If no indication of charge is shown, continue flight but place generator switch OFF.

EMERGENCY BATTERY OPERATION.

If failure of both APU's or both generators occurs, the emergency battery must be energized to power the battery bus. The battery bus supplies power for restarting the APU's or to control the APU's.

HYDRAULIC SYSTEM FAILURE.

No emergency system is provided for the hydraulically operated aerodynamic flight control system, speed brakes, and wing flaps. Some airplanes have an SAS emergency hydraulic system which automatically provides hydraulic power to the SAS pitch and roll servos in the event of a No. 2 hydraulic system failure.

FLIGHT CONTROL HYDRAULIC SYSTEM FAILURE.

If failure of one flight control hydraulic system occurs, the other system assumes the entire load of aerodynamic flight control system operation. (Refer to "Aerodynamic Flight Control System" in Section I.) However, under such a condition, available control surface hinge moments will be one half their design value, because of reduction of hydraulic flow.

ARTIFICIAL-FEEL SYSTEM FAILURE.

The artificial-feel system failure can be indicated by a lightening of stick forces (resulting in overcontrol), lack of trim response, and poor stick-centering characteristics. If failure of flight control artificial feel is encountered, proceed as follows:

1. Airspeed - Reduce.

   Reduction of airspeed may relieve severe oscillations of the airplane.

2. If control cannot be maintained - Eject.

   Ejection is recommended if control cannot be maintained. If partial control is available, thoroughly check flight characteristics before descending below safe ejection altitude.

TRIM SYSTEM FAILURE.

If the trim system fails in either extreme-travel position, the maximum force required by the pilot to move the control surface to the opposite extreme is not beyond physical capabilities. If the pitch trim knob fails, the trim system can be controlled by the alternate trim switch. The trim control switch must be moved to ALTERNATE before the alternate trim switch is operable.

STABILITY AUGMENTATION SYSTEM (SAS).

The stability augmentation system provides aerodynamic damping on all axes of the airplane. If any axis of the SAS fails, control can be maintained by the pilot.

Section III T.O. 1X-15-1

EMERGENCY ENTRANCE

EJECTION HANDLES NOT RAISED

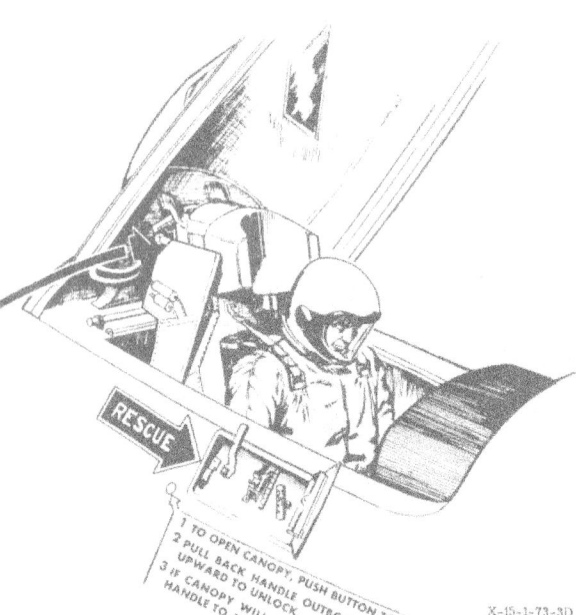

① Push yellow button to unlatch cover door on right side of forward fuselage.

② Pull out manual release handle and rotate handle up and back to unlatch canopy.

③ Lift canopy at forward end; then push canopy full open.

NOTE

If canopy cannot be opened with manual release handle, jettison canopy by pulling yellow "T" handle just forward of manual release handle.

WARNING

- There is no extension cable on the jettison handle.
- Keep all personnel clear of canopy ejection path.

④ Open helmet visor to ensure that pilot can breathe.

⑤ Cut ejection seat initiator hose at centerline just forward of pressure bulkhead.

NOTE

Hose cut point is painted red.

Figure 3-4 (Sheet 1 of 3)

6 Depress foot restraint release buttons if pilot's feet are locked in stirrups.

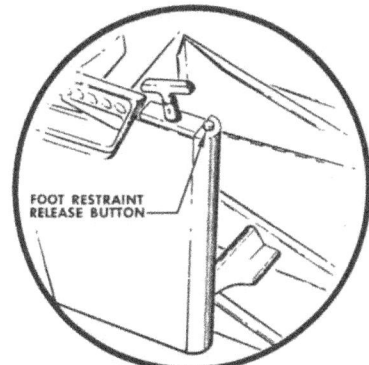

7 Disconnect personal leads.

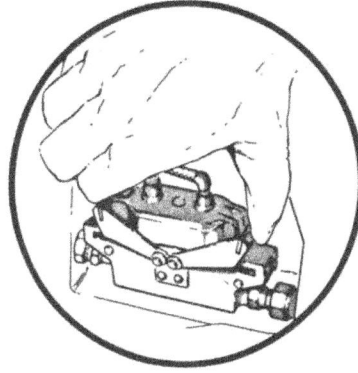

8 Open lap belt.

NOTE

If the lap belt is inside the sacrifice garment, the front zipper of the garment must be opened to reach the lap belt.

9 Disconnect parachute "D" ring retainer strap and both parachute risers, in that order.

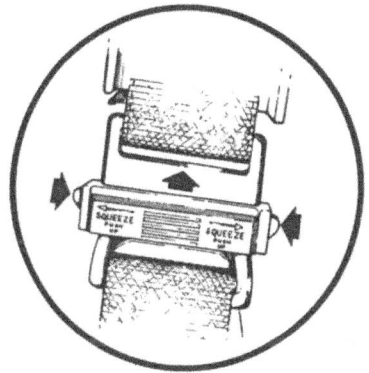

10 Remove pilot.

NOTE

If time does not permit manual release of the pilot, push canopy completely off the airplane and pull restraint emergency release handle to free pilot of all restraints. Pilot's parachute container will remain with pilot.

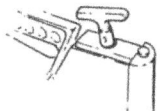

WARNING

Do not stand above seat when restraint emergency release handle is pulled, because the headrest is ejected straight up with explosive force.

X-15-1-73-5B

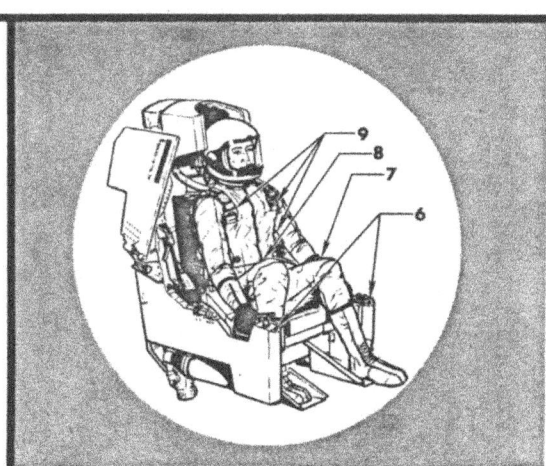

Figure 3-4 (Sheet 2 of 3)

Section III T.O. 1X-15-1

1 EMERGENCY ENTRANCE

WARNING

If the ejection handles are raised, this procedure must be followed. Jettisoning the canopy or manually removing the canopy before the seat is disarmed will cause the seat to eject.

EJECTION HANDLES RAISED

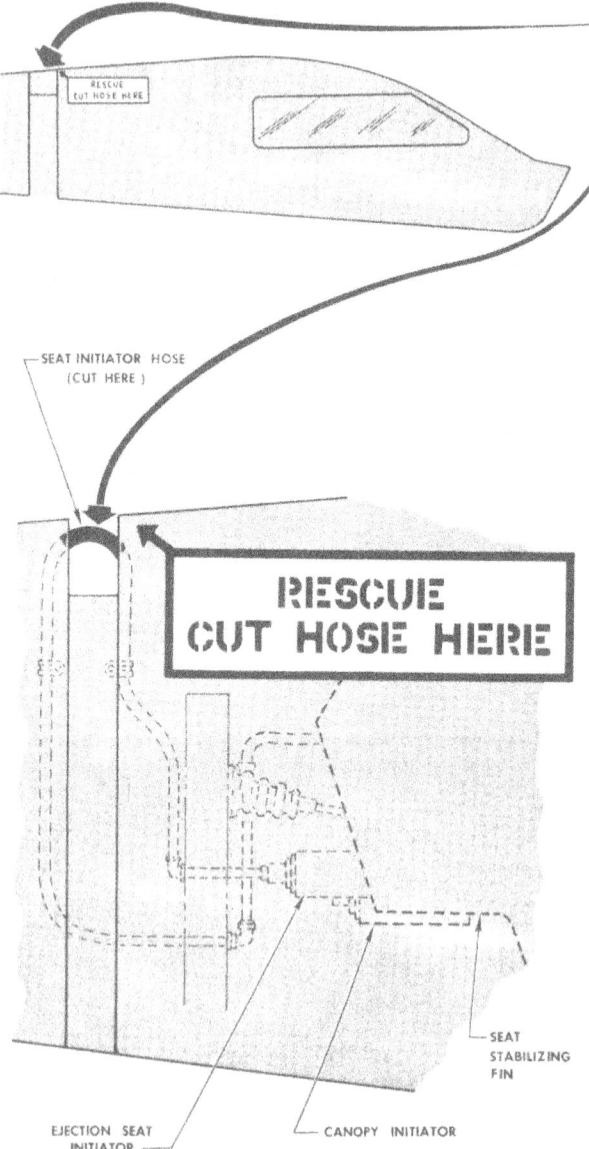

1 Remove access panel on top of fuselage, just aft of canopy.

2 Cut ejection seat initiator hose at centerline just forward of bulkhead.

NOTE
Hose cut point is painted red.

3 Push yellow button to unlatch cover door on right side of forward fuselage.

4 Pull out manual release handle and rotate handle up and back to unlatch canopy.

5 Lift canopy at forward end, then push up and aft until canopy completely separates from airplane.

NOTE
If canopy cannot be opened with manual release handle, jettison canopy by pulling yellow "T" handle just forward of manual release handle.

WARNING
- There is no extension cable on the jettison handle.
- Keep all personnel clear of canopy ejection path.

6 Open helmet visor to ensure that pilot can breathe.

7 Simultaneously push both ejection handles outboard and down to full stowed position.

8 Do steps 6 through 10 of "Ejection Handles Not Raised."

NOTE
If both ejection handles will not move full down, pull restraint emergency release handle to free pilot of all restraints. Then disconnect parachute "D" ring retainer strap and both parachute risers and remove pilot from seat.

WARNING
Do not stand above seat when restraint emergency release handle is pulled, because the headrest is ejected straight up with explosive force.

Figure 3-4 (Sheet 3 of 3)

EJECTION

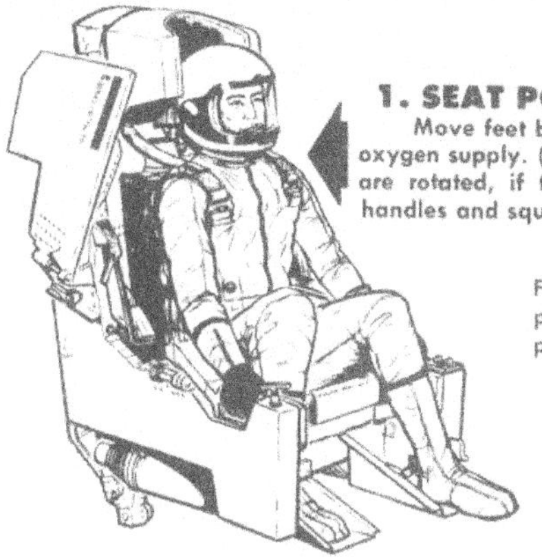

1. SEAT POSITION.

Move feet back and lock in foot restraints. Pull green ball for pilot's emergency oxygen supply. (The emergency supply is automatically actuated* when the handles are rotated, if time does not permit manual actuation.) Grip both seat ejection handles and squeeze either latch to release handles.

WARNING

Feet must be pulled full back into footrests to prevent injury by contact with the instrument panel or by wind blast.

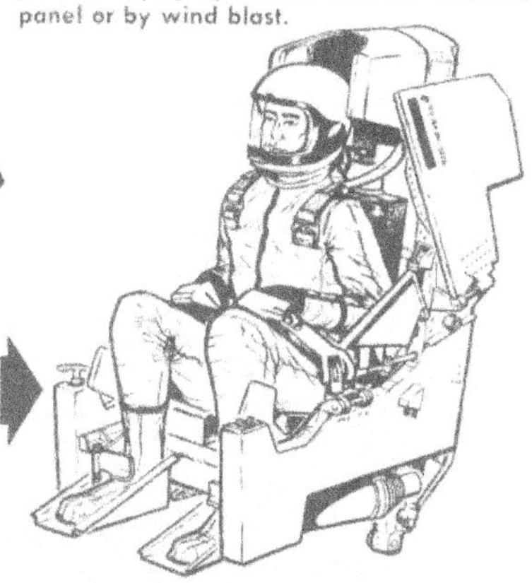

2. PREPARING TO FIRE.

Move head back firmly against headrest and pull seat ejection handles up and in toward chest until locked. The canopy will fire in the last 15 degrees of movement. After the canopy fires, the seat is automatically ejected.

3. SEAT FIRES.

After the seat catapult fires, and just before the seat leaves the ejection rails, a small rocket is automatically fired to increase the seat trajectory and aid in stability. Also at this time, the stabilizing boom and stabilizing fins are extended, and the restraint system aneroid is armed.

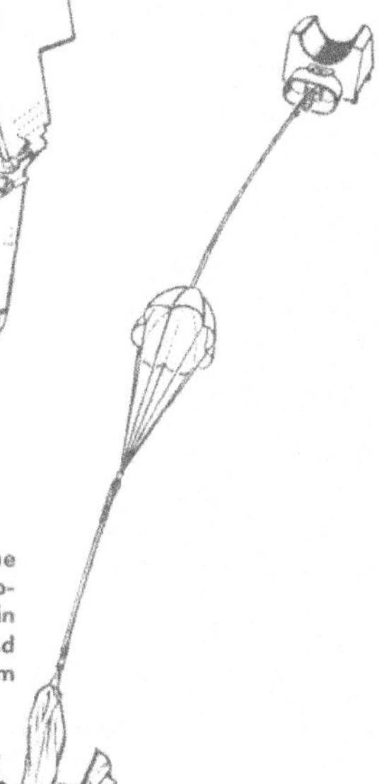

4. SEPARATION FROM SEAT.

Seat separation below 15,000 feet or after freefall down to 15,000 feet occurs in 3 seconds in the following sequence:

Releasing of the integrated restraint harness from the seat; unlocking of the seat ejection handles and foot restraint; and finally firing the headrest, which releases the shoulder harness, and deploys the pilot chute, which in turn deploys the main parachute to pull the pilot free of the seat.

* Some airplanes

Figure 3-5

BALLISTIC CONTROL SYSTEM FAILURE.

Failure of one system does not affect the operation of the other system, which is adequate to maintain attitude control. However, control will be at a reduced rate.

REACTION AUGMENTATION SYSTEM (RAS).

The reaction augmentation system provides pitch, roll, and yaw rate damping through the use of the No. 1 ballistic control system. If any axis of the RAS fails, attitude control is still maintained by the pilot. However, precise attitude control may deteriorate.

Auxiliary Equipment

SECTION IV

TABLE OF CONTENTS

	PAGE
Air Conditioning and Pressurization System	4-1
Communication and Associated Electronic Equipment	4-7
Lighting Equipment	4-9
Pilot's Oxygen System	4-10
Miscellaneous Equipment	4-10

AIR CONDITIONING AND PRESSURIZATION SYSTEM.

The air conditioning and pressurization system utilizes either liquid nitrogen or ram air as a cooling and pressurization medium. The liquid nitrogen is supplied from the X-15 storage tank during free flight and from either the X-15 storage tank or the carrier airplane pylon tank during captive flight. The X-15 tank can supply the entire air conditioning and pressurization system and when used will supply liquid nitrogen for pressurizing and cooling the cockpit and No. 2 electronic equipment compartment, and for cooling for the ac generators, APU upper turbine bearings, stable platform, and ball nose. It will also supply nitrogen gas for inflating the canopy and equipment compartment seals, purging the hydraulic reservoirs, between-panel purging of the windshield, and ventilation of the pilot's pressure suit. The nitrogen gas used for between-panel windshield purging and pilot's pressure suit ventilation is supplied through a common liquid nitrogen supply line which contains a heat exchanger and electric heater for converting the liquid to gaseous nitrogen. The carrier airplane pylon tank and associated controls make up the augmented cooling system. The primary function of this system is to augment the cooling to the stable platform during ground operation and taxiing, when the blowers are circulating compartment air, and when ram air is used during climb to cruise altitude. The augmented system is also used for between-panel windshield purging and pilot's pressure suit ventilation during these times and can be used for cooling the No. 2 electronic equipment compartment and ball nose.

The ram-air system is limited in its operation to cooling the cockpit and the electronic equipment compartment. The ram-air system, with augmented cooling, is used during climb and cruise. The X-15 liquid nitrogen system is turned on at the beginning of cruise-climb to launch altitude.

NOTE

If the ram-air system does not provide adequate cockpit cooling during captive flight at altitude, the augmented cooling system liquid nitrogen supply can be used to increase cooling by turning off the ram-air system and turning on the blowers.

The ram-air system should not be used above 35,000 feet or at speeds in excess of Mach 1.0, because of possible damage to the ram-air scoop and ducting due to high or excessive air loads and temperatures too high for proper electronic equipment cooling. During free flight, ram air may be used as an emergency measure if the X-15 liquid nitrogen system fails or during descent when the liquid nitrogen system has been depleted, but the limitation of Mach 1.0 must be observed. The cockpit and the electronic equipment compartment are pressurized to 3.5 psi by the use of nitrogen after it has been converted from liquid to gaseous state. The X-15 liquid nitrogen is stored under pressure and is converted as required for cockpit and equipment cooling and pressurization. A cockpit pressure regulator automatically maintains a pressure of 3.5 psi at 35,000 feet and above. From the ground up to 35,000 feet, the cockpit pressure remains the same as the outside pressure. If the cockpit pressure regulator does not operate properly, a cockpit safety valve will not allow the cockpit pressure to exceed 3.8 psi; also, during a rapid descent, the safety valve will prevent a pressure differential greater than 0.5 psi. Pilot comfort is provided by the MC-2 pressure suit, which maintains a regulated pressure of 3.6 psi over the entire body. Pressure to the suit is provided by the air conditioning and pressurization system. Temperature control of the nitrogen flow through the suit is maintained by an electric heater which is adjusted to a pilot-selected range. The temperature of the gas supplied to the pressure suit is from $55°F (\pm 5°F)$ to $90°F (\pm 5°F)$. The X-15 air conditioning and pressurization system may be considered automatic to the extent that once it has been started and manually regulated for desired flow and temperature, it will continue to operate without further adjustment.

COCKPIT AIR CONDITIONING AND PRESSURIZATION.

Cockpit air conditioning and pressurization is by either of two systems: the ram-air system, or the liquid nitrogen air conditioning and pressurization system of

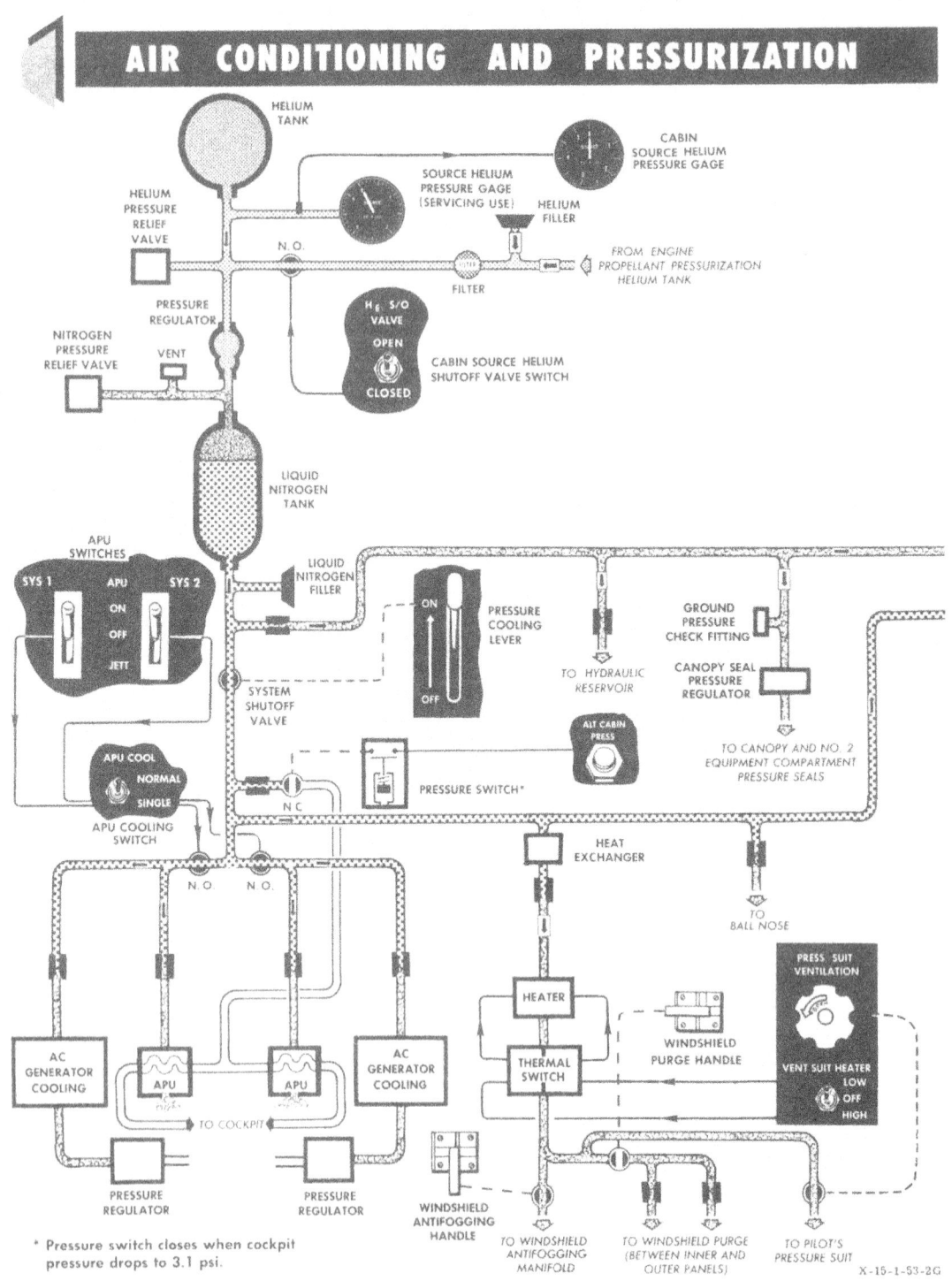

Figure 4-1

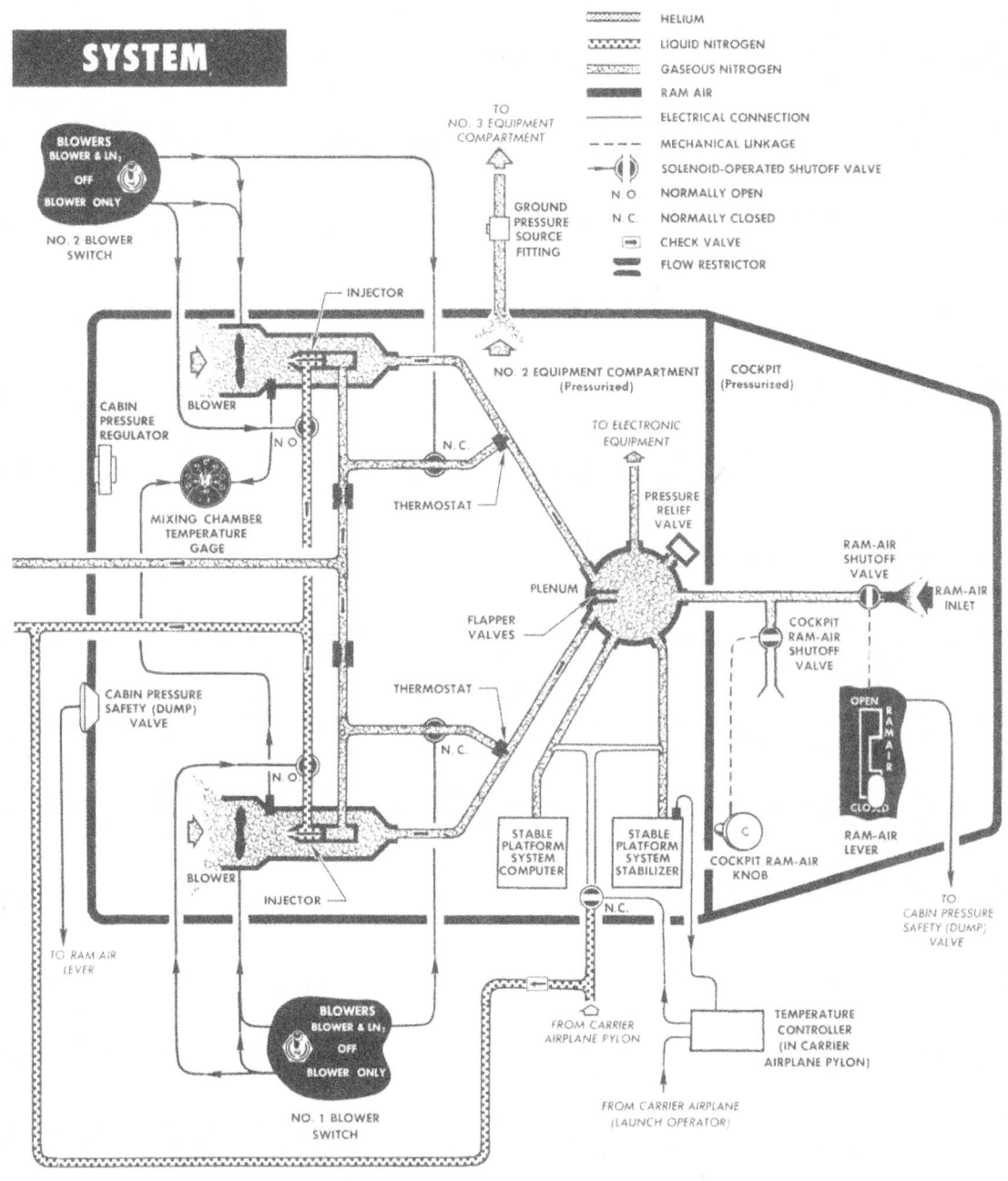

the X-15 Airplane. The ram-air system, although it does not pressurize the cockpit and the No. 2 electronic equipment compartment, will furnish adequate cooling for these areas. Approximately 10 percent of the ram-air flow may be diverted to the cockpit by a cockpit control, with the remaining 90 percent being routed through a plenum, from which it is ducted to the No. 2 electronic equipment compartment. Flapper-type shutoff valves prevent the ram-air flow from entering the mixing chambers and blower ducts of the cooling system. The ram-air system must not be operated at the same time as the X-15 nitrogen system. The X-15 liquid nitrogen system actually cools and pressurizes simultaneously. A vacuum-type contained in the No. 3 equipment compartment stores 176 pounds, or 26.5 gallons, of liquid nitrogen, which is depleted at a normal usage rate of approximately 5 pounds per minute. A pressure relief valve mounted in the nitrogen line between the top of the container and the nitrogen vent and build-up valve is preset to vent at 80 (+0, -4) psi. Helium gas for pressurizing the X-15 liquid nitrogen tank is stored in a spherical tank in the No. 3 equipment compartment. Helium pressure is reduced from approximately 4200 psi at the tank to approximately 67 psi through a differential-pressure regulator. The helium gas is then routed into the top of the liquid nitrogen tank. The liquid nitrogen is forced out of the bottom of the tank and on through the system to the injectors. The liquid nitrogen is forced under pressure through the injectors into a stream of returning gaseous nitrogen within the mixing chambers, where it is mixed and recirculated. The temperature of the air conditioning system is automatically controlled by the two thermostats that regulate the flow of nitrogen vapor from the liquid nitrogen injectors. The thermostats and the liquid nitrogen injectors are part of the system mixing chambers that connect with the plenum from which the regulated cooling gas is ducted to the electronic equipment requiring direct cooling. The air conditioning and pressurization system includes an alternate system which automatically pressurizes the cockpit if component malfunctions result in loss of pressure. The alternate system includes a pressure switch, a solenoid-operated shutoff valve, two heat exchangers, and associated plumbing. When the alternate system is armed and the pressure switch contacts close (cockpit altitude rises above approximately 37,000 feet), the solenoid-operated shutoff valve is energized open. This allows liquid nitrogen to flow to a heat exchanger in each APU compartment. The nitrogen is converted to a gaseous state and expelled into the cockpit. The pressure switch contacts open when cockpit altitude drops below approximately 36,000 feet. The augmented cooling system liquid nitrogen supply in the carrier airplane pylon is contained within a 6-cubic-foot tank. Approximately 292 pounds of liquid nitrogen is pressurized to approximately 63 psi by nitrogen pressure in the carrier airplane gaseous nitrogen system. Flow of the augmented cooling system liquid nitrogen to the X-15 is controlled from the launch operator's panel in the carrier airplane.

AIR CONDITIONING AND PRESSURIZATION SYSTEM CONTROLS AND INDICATORS.

Pressure-Cooling Lever.

The pressure-cooling lever (7, figure 1-4), on the right console, controls both the air conditioning and pressurization of the cockpit and the electronic equipment compartments. The lever has two positions, ON and OFF. When the lever is moved forward to ON, the system manual shutoff valve opens, allowing liquid nitrogen to flow throughout the entire system. With the lever at OFF, the system is inoperative. In free flight, the nitrogen gas for the pilot's pressure suit system is also controlled by the pressure-cooling lever. (Refer to "Pilot's Pressure Suit System" in this section.)

Alternate Cabin Pressurization Switch.

This push-pull, circuit-breaker type switch (5, figure 1-4), on the right console, controls primary dc bus power to the alternate cabin pressurization system pressure switch and to the system shutoff valve. The switch, labeled "ALT CABIN PRESS," is closed when pushed in. Pulling the switch out opens the alternate cabin pressurization system arming circuit.

Blower Switches.

There are two blower toggle switches (8, figure 1-4) on the circuit-breaker panel on the right console. The three-position switches, powered by the primary dc bus, are labeled "BLOWERS." When the switches are at OFF, the blowers are off, the liquid nitrogen shutoff valves to the injectors are de-energized open, and the pressure control valves are de-energized closed. With the switches at BLOWER & LN$_2$, electrical power is applied to the blowers, the liquid nitrogen shutoff valves to the injectors are de-energized open, and the pressure control valves are energized open. With the switches at BLOWER ONLY, electrical power is applied to the blowers, the liquid nitrogen shutoff valves to the injectors are energized closed, and the pressure control valves are de-energized closed. Both the No. 1 and No. 2 blowers are powered by the No. 1 primary ac bus. The No. 1 and 2 solenoid shutoff valves are powered by the primary dc bus. The blowers are protected by circuit breakers on the circuit-breaker panel.

Cabin Source Helium Shutoff Valve Switch.

Helium tank pressure to the liquid nitrogen tank is controlled by a two-position toggle switch on the circuit-breaker panel (6, figure 1-4) on the right console. The switch is labeled "CABIN SOURCE H$_e$ S/O VALVE." With the switch at CLOSED, primary dc bus power closes a solenoid-operated shutoff valve in the helium line upstream of the helium tank. With the switch at OPEN (forward), the solenoid-operated shutoff valve in the helium line is de-energized to the open position. The valve is used to isolate the air conditioning and pressurization system helium supply from the main propellant helium source as long as the former remains at or above 3000 psi. If the air conditioning and pressurization system helium supply drops below 3000 psi before launch, the valve should be opened to top off the helium supply.

Windshield Antifogging Handle.

This unlabeled handle is used to position a valve which controls flow of gaseous nitrogen to the inner surface of the windshield inner panels. It is on the right side of the canopy, just below the windshield panel and aft of the windshield purge handle. Flow of the nitrogen is shut off with the handle rotated down. Rotating the handle counterclockwise 90 degrees (to horizontal) opens the valve and permits flow of X-15 or augmented system gaseous nitrogen to the windshield antifogging manifold.

Windshield Purge Handle.

This unlabeled handle is used to position a valve which controls purging flow of gaseous nitrogen between the windshield panels. The handle is on the right side of the canopy, just below the windshield panel and forward of the windshield antifogging handle. Flow of the purging gas is shut off with the handle rotated down. Counterclockwise rotation of the handle 90 degrees (to horizontal) opens the valve and permits purging gas flow.

Ram-air Lever.

A ram-air lever (12, figure 1-5), on the left side of the center pedestal, controls the operation of the ram-air system. It is labeled "RAM AIR" and has OPEN and CLOSED positions. The lever is mechanically linked to the ram-air scoop and the ram-air shutoff valve, and electrically connected to the cockpit safety (dump) valve. When the lever is moved to OPEN, the ram-air shutoff valve and the ram-air scoop are opened to allow ram air to enter the system. At the same time, the cockpit safety valve is opened, allowing the cockpit and electronic equipment compartment pressure to be depleted and the ram air to be circulated through these areas. When the lever is at CLOSED, the ram-air shutoff valve and the ram-air scoop close. The cockpit safety valve is closed when the ram-air lever is at CLOSED. The ram-air scoop is on the lower centerline of the fuselage, just aft of the nose wheel well.

Cockpit Ram-air Knob.

Ram air to the cockpit is controlled by the cockpit ram-air knob (6, figure 1-5), on the center pedestal. The knob is mechanically linked to a shutoff valve that is ducted off the main ram-air line. When the knob is pulled straight out, the two-way shutoff valve opens to allow ram air (if the ram-air lever is at OPEN) to enter the cockpit. Pushing the knob in closes off the ram air to the cockpit.

The cockpit ram-air knob should be pushed in when the X-15 liquid nitrogen system is in use (blower switches at BLOWER & LN2); otherwise, recirculating cooling gas will be bled from the ducting system.

Cabin Helium Source Pressure Gage.

A cabin helium source pressure gage (35, figure 1-2), on the lower right of the instrument panel, indicates the helium pressure available to operate the air conditioning and pressurization system. The gage is calibrated from 0 to 5000 psi. The gage and pressure transmitters are powered by the 26-volt ac bus and protected by a circuit breaker on the circuit-breaker panel on the right console.

Mixing Chamber Temperature Gage.

The dual-pointer mixing chamber temperature gage (41, figure 1-2) shows the temperature in the No. 1 and No. 2 mixing chambers. Power is supplied by the No. 1 primary ac bus to amplifiers within the gage unit, activating the indicator needles. The gage shows temperature in degrees centigrade and is calibrated in 10°C increments from -80°C to +80°C.

Cabin Pressure Altimeter.

The pressure altitude of the cockpit is shown by a cabin pressure altimeter (33, figure 1-2) on the lower right corner of the instrument panel. The indicator is vented only to pressure within the cockpit, and operates on the aneroid principle.

OPERATION OF AIR CONDITIONING AND PRESSURIZATION SYSTEM.

Operation of Augmented Cooling System.

To operate the augmented cooling system, proceed as follows:

1. Cabin source helium shutoff valve switch - OFF.

2. APU cooling switch - SINGLE.

3. Pressure cooling lever - OFF.

4. Windshield antifogging handle - OFF.

5. X-15 LN$_2$ supply switch - Check OPEN.

Check with launch operator that switch is at OPEN.

6. Vent suit heater switch - HIGH or LOW.

7. Pressure suit ventilation knob - As required.

Knob should be open a minimum of one turn.

8. Stable platform LN$_2$ cooling control switch - Check OFF.

Check with launch operator that switch is OFF.

9. Ram-air lever - CLOSED.

10. Blower switches - BLOWERS & LN$_2$.

11. Mixing chamber temperature gage - CHECK.

When both chamber temperatures stabilize at -40°C (±5°C), proceed to step 12.

12. Stable platform switch - EXT.

Move stable platform switch to EXT and wait 10 minutes before proceeding to step 13.

13. Blower switches - BLOWER ONLY.

14. Stable platform LN$_2$ cooling control switch - Check ON.

Check with launch operator that switch is ON.

Operation of Ram-air and Augmented Cooling System (Captive Flight).

To operate the ram-air and augmented cooling systems, proceed as follows:

1. Cabin source helium shutoff valve switch - OFF.

2. APU cooling switch - SINGLE.

3. Pressure cooling lever - OFF.

4. X-15 LN$_2$ supply switch - Check OPEN.

Check with launch operator that switch is OPEN.

5. Stable platform LN$_2$ cooling control switch - Check OFF.

Check with launch operator that switch is OFF.

6. Blower switches - OFF.

7. Ram-air lever - OPEN.

Operation of X-15 Liquid Nitrogen System.

To operate the X-15 Airplane liquid nitrogen system at prelaunch, proceed as follows:

1. Ram-air lever - CLOSED.

2. Cabin source helium shutoff valve switch - CLOSED.

3. APU cooling switch - NORMAL.

4. Alternate cabin pressurization switch - ON (down).

5. Pressure-cooling lever - ON.

6. Vent suit heater switch - HIGH or LOW.

7. Pressure suit ventilation knob - As required.

8. X-15 LN$_2$ supply switch - Check CLOSE.

Check with launch operator that switch is at CLOSE.

9. Blower switches - BLOWER & LN$_2$.

NOTE

The windshield heater switches should be moved to ON (up) when fogging or frosting of the glass is observed and should remain ON (up) until the airplane is landed.

Emergency Operation of Ram-air System (After Launch).

To operate the ram-air system, proceed as follows:

1. Cabin source helium shutoff valve switch - OPEN.

2. Pressure-cooling lever - ON.

3. No. 1 and 2 blower switches - OFF.

4. Ram-air lever - OPEN.

NOTE

Do not operate the blowers and the ram-air system simultaneously.

WARNING

o Do not open the ram-air scoop when the airplane is above 35,000 feet or the airspeed is in excess of Mach 1.0; otherwise, the ram-air scoop will be subjected to excessive air loads and be damaged.

• If the ram-air lever is opened, cockpit pressure will be dumped through the cockpit safety valve.

GENERATOR AND APU COOLING AND PRESSURIZATION.

Nitrogen Cooling System.

The two ac generators and APU upper turbine bearing areas are cooled and pressurized by gaseous nitrogen after it has been converted from a liquid state. When the pressure-cooling lever and cockpit helium switch are ON, X-15 system liquid nitrogen flows to two normally open shutoff valves, one for each generator cooling shroud and each APU upper turbine bearing area. These valves are electrically connected to the APU switches so that when either APU switch is turned OFF while the APU cooling switch is at SINGLE, its respective valve closes. This shuts off the liquid nitrogen supply to the affected units regardless of the position of the pressure-cooling lever. After passing through the shutoff valves, the liquid nitrogen is routed to pressure restrictors, where it is changed to a gaseous state before moving on to the generators and turbines. The gaseous nitrogen, after it has absorbed heat from the generator and APU upper turbine bearing area, is exhausted into the APU compartment.

APU Nitrogen Cooling Switch. This two-position switch (3, figure 1-4), at the aft end of the right-hand console, is labeled "APU COOL." It is powered from the primary dc bus through the APU switches. With the switch at NORMAL, primary dc bus power is cut off from the shutoff valves which control liquid nitrogen flow to the APU's and ac generators, and the valves are de-energized open. With the switch at SINGLE, each valve will be energized closed if its respective APU switch is OFF, or de-energized open if its respective APU switch is ON. During single-APU operation, if excessive upper bearing temperatures are encountered on the operating APU, the switch should be moved to NORMAL. This will increase the volume of liquid nitrogen flow through individual flow restrictors, thus providing more efficient cooling of the bearing. If the upper bearing temperature of the nonoperating APU is excessively low, the switch should be moved to SINGLE, to shut off cooling nitrogen flow to the nonoperating APU.

CAUTION

If moving the switch to SINGLE results in excessive upper bearing temperatures on the operating APU, the switch must be returned to NORMAL, and if necessary, the nonoperating APU restarted.

APU Water Cooling System.

The water cooling system (figure 1-8) consists of an "APU COOL" switch, a water reservoir, a control valve, and the necessary plumbing to introduce water into the gaseous nitrogen cooling annulus surrounding the upper turbine bearing of each APU. The system is pressurized by turbine exhaust gas. Spent water is discharged into the APU compartment. The system

THIS PAGE
INTENTIONALLY
LEFT BLANK.

COMMUNICATION EQUIPMENT

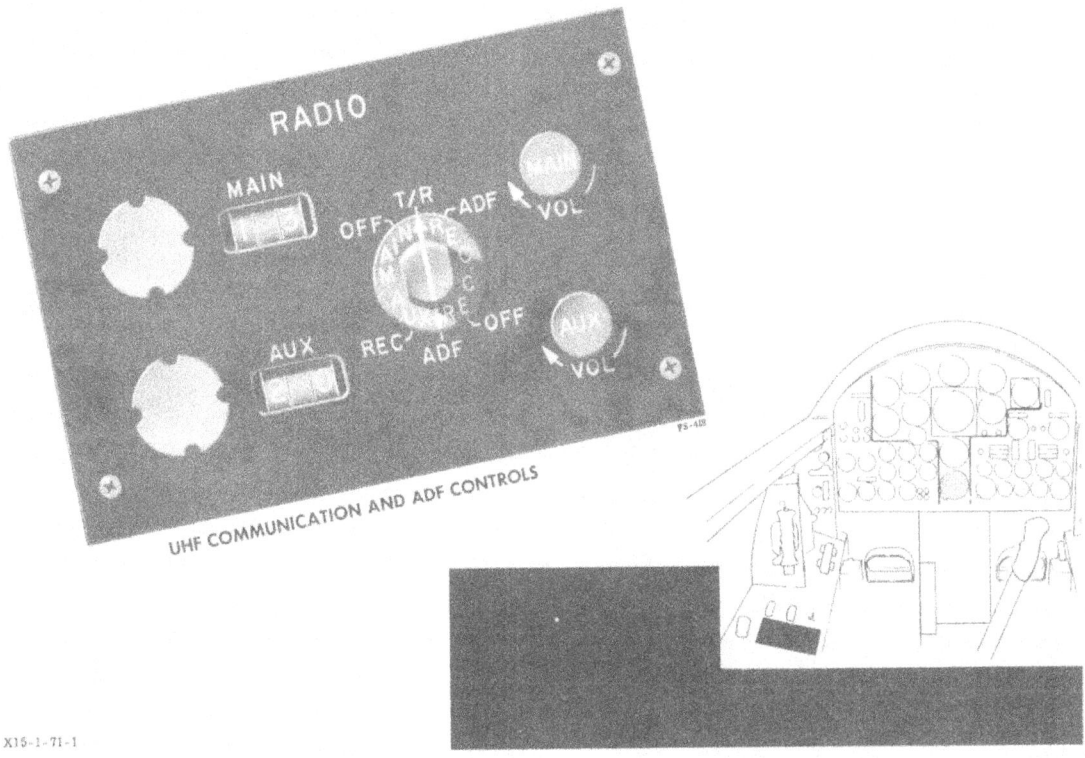

Figure 4-2

is designed to control excessive bearing temperatures during re-entry flight and is used at the pilot's option.

APU Water Cooling Switches. The two water cooling switches (64A, figure 1-2), mounted on lower left portion of the instrument panel, are labeled "APU COOL - 1" and " - 2," with ON and OFF positions. When a switch is at ON, primary dc bus power opens the APU water control valve to the respective APU, allowing water to flow to the cooling annulus. When the switch is at OFF, the valve is closed.

NOTE

The APU water cooling switches should not be confused with the nitrogen cooling switches on the right console. Both sets of switches are labeled "APU COOL."

WINDSHIELD HEATING SYSTEM.

Windshield frost or fogging is eliminated by a combination of heated nitrogen gas between the two glass panels and a heater element within each inner glass panel. A 1/4-inch line from the pilot's pressure suit supply line discharges a constant flow of heated nitrogen gas between the panels to dissipate moisture which may have collected in this area. Each heated inner panel is controlled by a toggle switch.

Windshield Heater Switches.

Two adjacent, two-position toggle switches, labeled "WINDSHIELD HEATER," control primary dc power to the windshield heater elements, one for each side panel. The switches are located in the lower right corner of the instrument panel right wing. (See 29, figure 1-2.) When the switches are moved ON (up), primary dc bus power is applied to the heater elements.

COMMUNICATION AND ASSOCIATED ELECTRONIC EQUIPMENT.

UHF COMMUNICATION SYSTEM.

The UHF communication system enables the pilot to transmit and receive voice transmissions. The system has two modes of operation, normal and alternate. During normal operation, the UHF transmitter and

Section IV   T.O. 1X-15-1

main receiver ("MAIN") are used with the UHF antenna for communication. The alternate "AUX" mode is used in case one of the receivers fails and it is desired to regain the function of the inoperative receiver. During this mode of operation, the UHF transmitter is connected to the UHF antenna through an automatic lockout feature for communication reception. The transmitter and main receiver cover the frequency range of 255.0 to 399.9 megacycles for 1750 channels, spaced at 100-kilocycle increments.

NOTE

Frequencies below 265 megacycles should not be used, because they are reserved for telemetering channels and the system will not accommodate this frequency range.

The guard receiver, an integral part of the system, operates off the UHF antenna and guards the preset frequency of 279.9 megacycles. The guard receiver has two functions: it operates the automatic antenna selector, and it also provides for reception of voice transmission of the guard frequency (normal guard operation). The auxiliary receiver has 20 channels in the frequency range of 265.0 to 284.9 megacycles, with one channel per megacycle increments. The system has a dual power supply, which gives the equipment continuous operation if one of the power sources fails. The main power source for the system is the three-phase ac and the 28-volt dc systems, and protection is afforded by circuit breakers. Controls for operation of the UHF communication system are on the radio control panel (figure 4-2), on the left console.

UHF Communication Controls.

Function Selector Switch. A rotary-knob type selector switch, on the radio control panel, controls the mode of operation, either normal or alternate. The switch is marked "MAIN REC" and "AUX REC" and has a two-headed arrow across the full diameter of the rotary knob. On the outer perimeter of the knob is the position marking denoting the function of the system. OFF, T/R and ADF are the positions for the main receiver; REC, ADF, and OFF are the positions for the auxiliary receiver. The ADF positions are inoperative in this airplane. The switch is designed so that when the arrow indicates OFF on the main system, the opposite arrow is indicating OFF on the auxiliary system. With the switch in the T/R position on the UHF receiver-transmitter, the auxiliary receiver is at the ADF position and is disconnected from the antenna. This is the normal mode of operation, and the main receiver is connected to the UHF antenna for communication. In the alternate mode of operation, the auxiliary receiver is connected to the UHF antenna for reception of UHF voice transmission. The functions of the main and auxiliary receivers are the same, provided the main and auxiliary channel selector knobs are set to the channel that corresponds to the frequency of the desired UHF signals.

NOTE

The main channel selector control sets up the frequency for the transmitter as well as the receiver.

Volume Controls. There are two volume controls on the radio control panel, marked "MAIN" and "AUX." When either of these rotary volume controls is rotated in the direction indicated by the arrow (clockwise), the volume of the respective equipment to the pilot's headset is increased. Counterclockwise turning of either volume will diminish the volume of the equipment being adjusted.

Channel Selector Knobs. There are two rotary selector knobs on the radio control panel used in selecting the desired channel frequency of the equipment in use. The UHF transmitter and main receiver channel selector control provides selection of 20 preset channels within the frequency range of the UHF receiver-transmitter. A similar channel control selector enables the pilot to select one of 20 channels on the auxiliary receiver.

Channel Indicators. There is a channel indicator window to the right of each channel selector knob. The upper window, marked "MAIN," shows the channel number that has been selected for the UHF transmitter and main receiver. A similar window, marked "AUX," to the right of the auxiliary receiver channel selector knob, displays the channel selected for the auxiliary receiver. (Deleted)

Changed 26 April 1963

Operation of UHF Communication System.

Normal Operation. The normal operating procedure is as follows:

1. Function selector switch - T/R (main system).

2. Main channel selector knob - Select desired channel.

3. Auxiliary channel selector knob - Select desired channel.

4. (Deleted)

5. "MAIN" and "AUX" volume controls - As desired.

Alternate Operation. If the main receiver fails, the auxiliary receiver is used. To use the auxiliary receiver, proceed as follows:

1. Function selector switch - REC (auxiliary receiver).

2. Main channel selector knob - Select desired channel.

3. Auxiliary channel selector knob - Select desired channel.

4. "MAIN" and "AUX" volume controls - As desired.

RADAR BEACON SYSTEM.

Radar beacon system ED-I-519 automatically receives and transmits radio-frequency signals from a ground station to enable ground radar tracking of the X-15 Airplane. The receiver and the transmitter of the radio beacon are tuned independently to a preselected frequency within the 2700- to 2900-megacycle range. The beacon is designed to operate dependably from a minimum range of 1200 yards to a line-of-sight range of at least 150 miles when operated in conjunction with a ground radar set, SCR-784 or equivalent. The antenna for the radar beacon is on the bottom of the fuselage, just forward of the UHF antenna. The radar beacon receiver and transmitter use the same antenna. The system is powered by the No. 1 primary ac bus through a switch in the cockpit.

Radar Beacon Switch.

The radar beacon switch (11, figure 1-5), on the center pedestal, controls No. 1 primary bus power to the radar beacon system. The switch has two positions, ON and OFF.

INTERCOMMUNICATION SYSTEM.

The X-15 pilot's microphone button and headset can be connected to the carrier airplane intercommunication system through a switch in the cockpit. Depending on carrier airplane intercommunication system switch settings, the X-15 pilot can communicate directly with the carrier airplane or transmit and receive by means of the carrier airplane command radio.

Intercommunication Switch.

The two-position intercommunication switch (2, figure 1-3), on the left console and labeled "INTERCOM," controls the function of the X-15 pilot's microphone buttons and headset. With the switch at ON, the X-15 pilot can communicate with the carrier airplane through the carrier airplane intercommunication system or transmit and receive command radio signals through the carrier airplane command radio, depending on the position of the carrier airplane intercommunication system function selected. With the switch at OFF, the X-15 pilot's microphone buttons and headset are connected to the X-15 command radio.

NOTE

If the intercommunication switch is at ON at time of launch, the pilot will not be able to use the command radio. This will be readily noticeable by the lack of static background noises or side-tone signals in the headset.

LIGHTING EQUIPMENT.

COCKPIT LIGHTING SYSTEM.

The cockpit is lighted by two floodlights attached to the canopy and located so that the light is directed forward and down, covering the entire cockpit controls and gages. These adjustable lights are shielded to eliminate direct glare to the pilot. A switch on the right instrument console controls the lighting system. The lighting system is powered by the primary dc bus.

NOTE

Loss of the canopy in flight will cause loss of the cockpit lighting system.

Cockpit Lighting Switch.

A two-position switch (32, figure 1-2), on the instrument panel right wing, controls the cockpit lighting system.

The switch has ON and OFF positions and is labeled "COCKPIT LIGHTS." The switch is powered by the primary dc bus.

PILOT'S OXYGEN SYSTEM.

The pilot's oxygen system is a high-pressure (3000 psi), gaseous-type system, completely contained within the ejection seat and pilot's back pan. Two 96-cubic-inch, lightweight cylinders are mounted on the bottom of the seat. The charging valve for both cylinders is about 6 inches above the seat bucket on the back rest. This system will supply 100 percent oxygen to the helmet for approximately 45 minutes at a pressure of 1-1/2 inches of water above the pilot's suit pressure, which varies with altitude. A reducer valve, incorporated in the high-pressure line, reduces the cylinder pressure from 3000 psi to 70 to 90 psi at the pilot's personal-lead quick-disconnect. The oxygen then flows into the back pan through a regulator and into the face mask. The system includes a warning system to indicate when breathing oxygen pressure is approaching the minimum for which the pressure suit is designed to permit normal breathing. No dilution is required in this system. Breathing oxygen is supplied by the carrier airplane from take-off until just before launch, at which time the pilot selects the X-15 Airplane oxygen supply. The oxygen supply is also utilized to pressurize the pilot's pressure suit in case the nitrogen ventilation and pressurization supply should fail.

WARNING

The oxygen supply to the pressure suit helmet must be on and the helmet visor must be down before the cockpit is pressurized with nitrogen.

If ejection is necessary, the complete oxygen system is retained with the seat. Immediately before ejection, the green ball on the upper right side (chest area) of the pressure suit should be pulled to activate the emergency oxygen system and maintain suit pressurization after separation from the seat. When the green ball is pulled, the oxygen in the back pan flows through a pressure reducer, on through an oxygen regulator and a pressure suit regulator, to the pressure suit and helmet. The emergency oxygen supply, contained in the back pan, has a capacity of about 100 cubic inches at a cylinder pressure of 1800 psi and for a duration of about 20 minutes. During the time interval between seat separation from the airplane and pilot separation from the seat, the breathing oxygen is drawn from the normal system oxygen bottles attached to the seat. This is made possible by a valve in the emergency system, which is kept closed by a pressure differential until the pilot separates from the seat.

PILOT'S OXYGEN SYSTEM SELECTOR AND PRESSURE GAGE.

A three-position oxygen system selector, mounted on the forward end of the left side panel of the seat (figure 1-14), allows pilot selection of two separate oxygen systems. The three positions, labeled "B-52," "OFF,"

and "X-15," depend on the phase of flight as to which system is selected. The control head is a round, tubular aluminum section with the oxygen pressure gage mounted in the center. The selector is positioned by turning it from the middle OFF position, either right or left, to select the desired system. The gage range is from 200 to 3000 psi. Permanent markings are on the range scale. There is a red radial at 200 and a green arc from 200 to 3000. If the breathing oxygen is depleted to the point where the gage pointer is at 200, the green ball on the pilot's pressure suit must be pulled to actuate the emergency breathing oxygen supply.

PILOT'S OXYGEN-LOW CAUTION LIGHT.

The pilot's oxygen-low caution light (3A, figure 1-2) is on the instrument panel, directly above the subsonic airspeed indicator. The placard-type amber light is powered by the primary dc bus and when illuminated reads "PILOTS O_2 LOW." It comes on when breathing oxygen pressure downstream of the reducer valve drops to about 60 psi. When the light comes on, the pilot is alerted to the possibility of encountering oxygen supply pressure sufficiently low to ultimately result in breathing difficulty.

NORMAL OPERATION OF PILOT'S OXYGEN SYSTEM.

For operation of the oxygen system, proceed as follows:

1. Upon entry into cockpit, move oxygen system selector to B-52.

2. Before launch, move oxygen system selector to X-15.

EMERGENCY OPERATION OF PILOT'S OXYGEN SYSTEM.

If the pilot's oxygen-low caution light comes on after the X-15 oxygen system has been selected, be prepared to actuate the emergency oxygen supply as soon as you are aware of breathing difficulty. The time between illumination of the light and when breathing difficulty is first encountered is dependent on several factors, including breathing rate, pilot size, and cabin pressure altitude.

MISCELLANEOUS EQUIPMENT.

PILOT'S PRESSURE SUIT SYSTEM.

The pilot's full-pressure suit is ventilated and pressurized by either the X-15 or the augmented cooling system liquid nitrogen supply. The liquid nitrogen flows through a heat exchanger and flow restrictor, where it is converted to gaseous nitrogen, then through a check valve, an electrical heater, and then to the vent suit flow control valve. From the vent suit flow control valve, the gaseous nitrogen is routed through a flexible disconnect hose and coupling to the pressure suit. After launch, ventilation of the pressure suit is maintained by the nitrogen system of the X-15 Airplane. Suit pressure is maintained at 1/10 psi above cockpit pressure at all altitudes above 35,000 feet. If cockpit pressure should fail, the nitrogen supply will pressurize the suit to maintain the 35,000-foot environment. The same routing lines and controls are used whether the nitrogen comes from the carrier airplane or the X-15. The desired volume of flow can be controlled from the cockpit. Temperature of the gaseous flow to the pressure suit may be controlled from approximately 50°F to 90°F by the vent suit gaseous heater. The heater is of the electric-tubular type and is controlled from the cockpit. Pressure and ventilation to the pressure suit are maintained automatically up to the vent suit flow control valve, when the air conditioning and pressurization system is in operation, by either the carrier airplane or the X-15 nitrogen system. The pressure suit is equipped with a pressure regulator that is used for suit ventilation, and an anti-blackout pressure regulator that operates from the breathing oxygen supply. In addition to these provisions, the suit back pan also contains a demand-type oxygen regulator for normal and emergency use and a stored supply of oxygen to be used during emergency ejection. An integrated parachute harness and restraint assembly also forms a part of the pressure suit.

Pressure Suit Ventilation Knob.

A knob (5, figure 1-3), on the left console, controls the flow of gaseous nitrogen to the pressure suit. This knob, marked "PRESS. SUIT VENT," has an arrow to indicate the direction the knob should be turned (counterclockwise) to increase the volume and flow of the nitrogen to the suit. The knob has direct mechanical linkage to the flow control valve. Nitrogen flow is restricted to 7 cfm by a stop on the knob. This is to prevent excess nitrogen flow which could override the capabilities of the vent suit heater in addition to causing suit overpressurization. The 7 cfm provides sufficient ventilating gas flow for normal conditions. When initiating suit ventilation, the knob should be opened about one turn and then the vent suit heater switch should be moved to either HIGH or LOW, as required. If additional ventilation is required, the knob can be opened further.

Vent Suit Heater Switch.

The vent suit heater switch (6, figure 1-3), on the left console, is a three-position switch. With the switch in the HIGH position, the gaseous nitrogen to the vent suit is heated to a temperature of approximately 75°F to approximately 90°F and is maintained within that range by a thermoswitch. With the switch in the LOW position, another thermoswitch maintains a temperature of approximately 55°F to approximately 70°F. With the switch in the OFF position, the heater is inoperative, and the temperature of the gaseous nitrogen is that of the system temperature.

The heater switch is powered by the primary dc bus. The heater is powered by the primary ac busses.

| CAUTION |

- When the augmented cooling system liquid nitrogen supply is being used during captive flight, the vent suit heater switch should be at HIGH or LOW, as necessary, to prevent liquid nitrogen from entering the area between the inner and outer windshield panels and consequently obscuring vision.

- When either the augmented cooling system or the X-15 liquid nitrogen system is supplying gaseous nitrogen to the pilot's ventilated suit, the vent suit heater switch should be at HIGH or LOW, as necessary, to ensure that the liquid nitrogen is converted to gaseous nitrogen.

- When there is no nitrogen flow (augmented cooling and X-15 liquid nitrogen systems turned off), the heater switch should be turned off to prevent damaging the heaters and equipment adjacent to the heaters.

Face Mask Heater Switch.

Primary dc bus power for the helmet visor heat elements is controlled by a four-position switch (3, figure 1-3), on the left console. The switch is labeled "FACE MASK HTR." The face mask heater is energized when the switch is moved from OFF. Heat is increased as the switch is moved from the LO to the MED and HI positions.

Face Mask Heater Alternate Switch.

Power for alternate heating of the face mask is provided from a battery on the ejection seat and is controlled by the face mask heater alternate switch, labeled "ALT DEFOG," on the forward end of the seat (figure 1-14). The face mask heater is energized when the switch is moved from the center OFF position. Heat increases when the switch is moved left to HI and decreases when moved to the right to LO.

Photographic Equipment.

High-altitude research cameras are installed on modified airplanes. The control circuit is armed whenever APU's are turned on before launch and set in motion automatically at launch.

CHANGED 26 APRIL 1963

THIS PAGE INTENTIONALLY LEFT BLANK.

T.O. 1X-15-1

Operating Limitations

SECTION V

TABLE OF CONTENTS	PAGE		PAGE
Instrument Markings	5-1	Acceleration Limitations	5-7
Engine Limitations	5-1	Roll Limitations	5-7
Captive Flight Limitations	5-1	Yaw (Sideslip) Limitations	5-7
Launch Limitations	5-1	Prohibited Maneuvers	5-7
Airspeed Limitations	5-1	Ballistic Control System Limitations	5-7
Ventral Jettison Limitations	5-6	Landing Limitations	5-7
Speed Brake Operation	5-6	Center-of-Gravity Limitations	5-7
Stability Augmentation System Limitations	5-7	Weight Limitations	5-7

INSTRUMENT MARKINGS.

Careful attention must be given to the instrument markings (figure 5-1), because the limitations shown on these instruments and noted in the captions are not necessarily repeated in the text of this or any other section.

ENGINE LIMITATIONS.

All normal engine limitations are shown in figure 5-1 and are based on the use of oxidizer and fuel specified in figure 1-16.

CAPTIVE FLIGHT LIMITATIONS.

Control surfaces may be deflected only individually. Full throw of the vertical stabilizer or flaps is permissible up to 300 knots EAS. The horizontal stabilizer may be deflected full throw at 200 knots EAS, decreasing linearly to one-half throw at 300 knots EAS. To avoid possible buffet on both the X-15 and carrier airplane, the speed brakes should not be extended beyond 10 degrees.

NOTE

Transient control surface deflections which will occur during captive flight checks of the MH-96 system, installed on Airplane AF56-6672, are within captive flight limitations.

LAUNCH LIMITATIONS.

Launch from the carrier airplane should be made under the following conditions:

 a. Mach .75 at 35,000 feet to Mach .82 at 45,000 feet.

 b. Carrier airplane in 1 G symmetrical flight.

 c. X-15 control surfaces as follows:

 Vertical stabilizer-ventral neutral.

 Horizontal stabilizer neutral in roll and 0 degrees to -2 degrees (leading edge down) in pitch. The setting of -2 degrees should not be used until some familiarity with launch characteristics has been gained at the 0-degree setting.

 d. Launch should not be made with the ventral jettisoned or not installed, except when the scheduled mission specifically requires it.

AIRSPEED LIMITATIONS.

MAXIMUM ALLOWABLE AIRSPEEDS.

The maximum allowable Mach number for varying flight altitudes is shown in figure 5-2. The maximum q (dynamic pressure) is 2200 psf with modified movable vertical stabilizer (Part No. 240-23001-201) and ventral (Part No. 240-23501-101 or 240-23502-201).

WARNING

Because of the exploratory nature of the X-15 research program, limitations for all flights beyond 3.5 Mach number will be based on separate mission analysis.

LANDING GEAR AND WING FLAP LOWERING SPEED.

Maximum allowable airspeed for extending the landing gear and wing flaps, or for flight with the landing gear and wing flaps fully extended, is 300 knots IAS. Flight with the landing gear and flaps extended at speeds above 300 knots IAS may cause structural damage.

CHANGED 29 JUNE 1962

Section V T.O. 1X-15-1

INSTRUMENT MARKINGS

ENGINE INSTRUMENTS, AIRSPEED INDICATOR, AND ACCELEROMETER

PROPELLANT SOURCE PRESSURE GAGE

NO MARKINGS

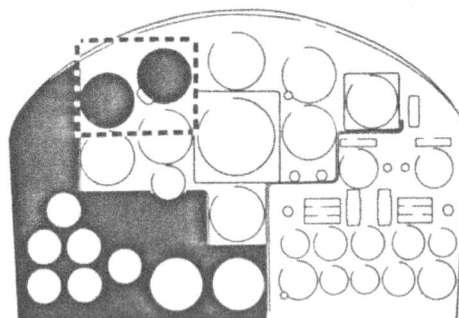

PROPELLANT PUMP INLET PRESSURE GAGE

PROPELLANT TANK PRESSURE GAGE

- 45-63 psi — Normal
- 70 psi — Maximum

- 30 psi — Minimum
  Ammonia pressure below 30 psi will require engine operation at reduced thrust. [Refer to "Fuel Line Pressure Low (XLR99 Engine)" in Section III.]
- 39-63 psi — Normal
- 75 psi — Maximum

PROPELLANT MANIFOLD PRESSURE GAGE

- 455-1155 psi — Normal
  (Readings as low as 350 psi are normal for 40% thrust configuration.)

H₂O₂ SOURCE AND PURGE PRESSURE GAGE

NO MARKINGS

H₂O₂ TANK AND ENGINE CONTROL LINE PRESSURE GAGE

- 445-psi — Minimum
- 550-615 psi — Normal
- 775 psi — Maximum

X-15-1-51-7D

Figure 5-1. (Sheet 1 of 4)

CHANGED 26 APRIL 1963

T.O. 1X-15-1　　　　　　　　　　　　　　　　　　　　　　　　　　　　Section V

AIRSPEED INDICATOR

■ 300 knots IAS maximum gear-and-flaps-down limit airspeed

ACCELEROMETER

Refer to "Acceleration Limitations" in this section for acceleration limitations at various airplane gross weights and flight conditions.

CHAMBER AND STAGE 2 IGNITER PRESSURE GAGE
MAIN CHAMBER (LONG POINTER)

■ 335-600 psi　Normal (Readings as low as 265 psi are normal for 40% thrust configuration.)
■ 630 psi　Maximum

LIQUID OXYGEN BEARING TEMPERATURE GAGE

■ −40°C　Minimum　If bearing temperature is below this value, launch must be aborted.
■ −29°C to 38°C　Normal

Figure 5-1. (Sheet 2 of 4)

CHANGED 26 APRIL 1963　　　　　　　　　　　　　　　　　　　　　　　　　5-3

Section V
T.O. 1X-15-1

INSTRUMENT MARKINGS

ELECTRICAL, HYDRAULIC, AND COCKPIT INSTRUMENTS

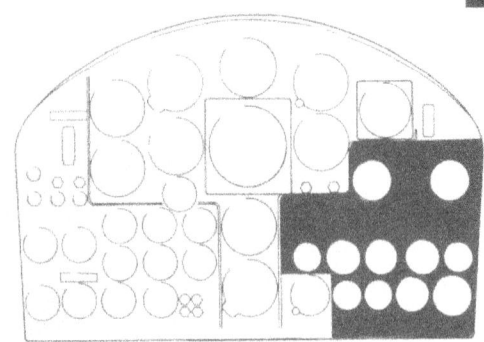

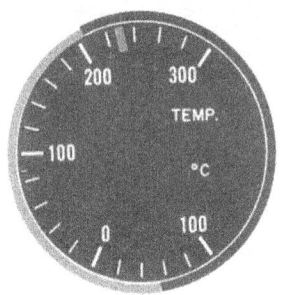

HYDRAULIC SYSTEM NO. 1 TEMPERATURE GAGE

- −54°C TO +204°C NORMAL
- 230°C MAXIMUM

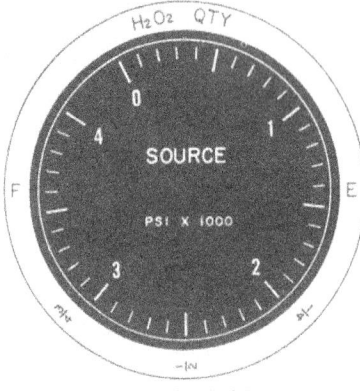

APU SOURCE PRESSURE GAGE

NO MARKINGS

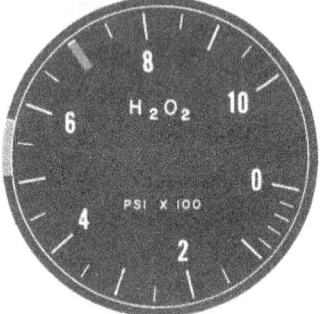

APU HYDROGEN PEROXIDE TANK PRESSURE GAGE

- 510-575 PSI NORMAL
- 700 PSI MAXIMUM

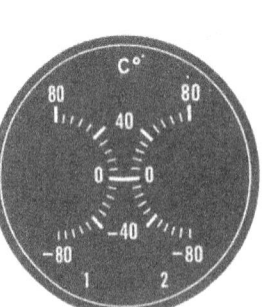

MIXING CHAMBER TEMPERATURE GAGE

NO MARKINGS

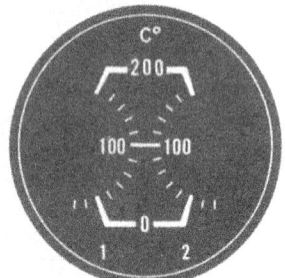

APU BEARING TEMPERATURE GAGE

NO MARKINGS

X-15-1-51-3C

Figure 5-1. (Sheet 3 of 4)

5-4

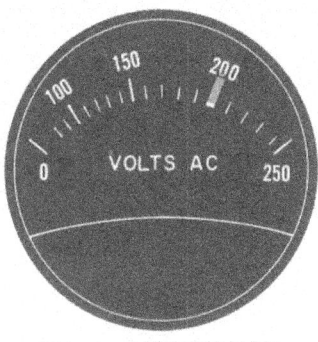

NO. 1 GENERATOR VOLTMETER

200 VOLTS NORMAL

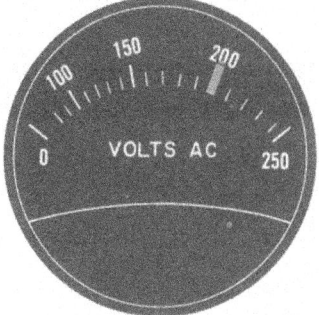

NO. 2 GENERATOR VOLTMETER

200 VOLTS NORMAL

HYDRAULIC PRESSURE GAGE

2900-3400 PSI NORMAL
3500 PSI MAXIMUM
NOTE

For a period not to exceed 30 seconds after APU start, pressure exceeding 4000 is permissible.

HYDRAULIC SYSTEM NO. 2 TEMPERATURE GAGE

−54°C TO +204°C NORMAL
230°C MAXIMUM

CABIN HELIUM SOURCE PRESSURE GAGE
NO MARKINGS

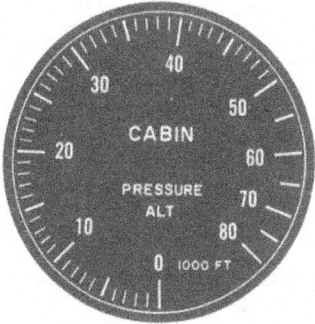

CABIN PRESSURE ALTIMETER
NO MARKINGS

Figure 5-1. (Sheet 4 of 4)

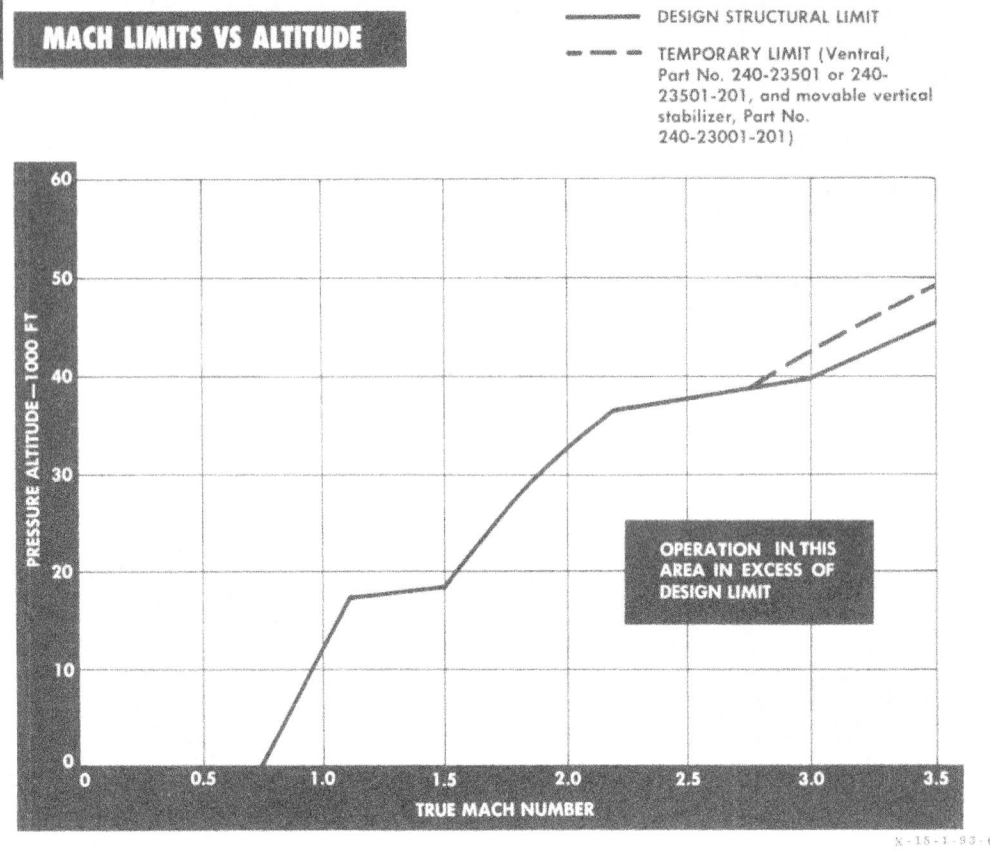

Figure 5-2.

VENTRAL JETTISON LIMITATIONS.

For all flights where the ventral is installed for launch, the ventral should not be jettisoned at speeds above 300 knots IAS or Mach 3.5, whichever is lower. Under normal flight conditions, with the SAS functioning properly, the ventral should not be jettisoned except during the landing approach.

NOTE

In the event of controllability problems arising from the loss of various SAS modes (primarily loss of the SAS roll mode), jettisoning of the ventral greatly improves airplane handling characteristics in the Mach number range of 2.0 to 3.5.

With the ventral jettisoned, the following limits apply:

Maximum angle of attack, 16 degrees

Maximum rate of roll, 30 degrees per second

Maximum Mach number, 3.5

SPEED BRAKE OPERATION.

The upper and lower speed brakes must be operated simultaneously and symmetrically. This restriction is imposed because no analysis has been conducted to determine limitations required for asymmetrical operation of the speed brakes.

CAUTION

Be cautious when using speed brakes for approach control, because rate of descent approximately doubles with the speed brakes open.

NOTE

Normally, the speed brake handles are mechanically interconnected to ensure simultaneous and symmetrical operation of the speed brakes.

STABILITY AUGMENTATION SYSTEM LIMITATIONS.

For normal flight operations, SAS gain settings should be:

Ventral on - pitch, 8; roll, 6; yaw, 8.

Ventral off - pitch, 8; roll, 4; yaw, 8.

Under conditions of a single system APU failure or failure of either the roll or yaw functions, lateral-directional control problems can be encountered in the speed range of Mach 2.5 to 6.0 and at angles of attack in excess of 8 degrees nose up with the ventral installed. Should control problems be encountered under these conditions, the following recovery techniques should be employed:

1. Reduce angle of attack.

2. Extend speed brakes.

3. Jettison ventral if below the airspeed limit for jettisoning the ventral.

NOTE

If the ventral is not installed, the failures noted previously will not seriously affect airplane controllability.

ACCELERATION LIMITATIONS.

In order to ensure correct interpretation of the two basic types of acceleration limitations imposed on this airplane, the following definitions apply:

a. Symmetrical flight is that where airplane bank angle is constant.

b. Asymmetrical flight is that involving a bank angle change rate, i.e., rolling flight.

SYMMETRICAL.

Acceleration limitations for symmetrical flight at various gross weights are shown graphically in figure 5-3.

Negative G imposed at low engine propellant levels may result in premature main chamber burnout.

ASYMMETRICAL.

Acceleration limitations for various gross weights from 33,000 to 19,000 pounds are shown in figure 5-4. At burnout gross weight (15,119 pounds), asymmetrical limitations are 5.2 G and 0 G.

ROLL LIMITATIONS.

The maximum allowable rate of roll is 100 degrees per second. Roll coupling can be encountered at the higher Mach numbers and angles of attack. Abrupt aileron rolls under these conditions should be simulator-tested before flight.

YAW (SIDESLIP) LIMITATIONS.

Abrupt yawing maneuvers utilizing maximum rudder rate and deflection should not be performed. Yaw maneuvers exceeding one half of the maximum rudder deflection in one direction, followed immediately by a similar deflection in the opposite direction should not be performed. The maximum permissible yaw (sideslip) angles at varying Mach numbers and altitudes are shown in figure 5-5.

PROHIBITED MANEUVERS.

The airplane is restricted from performing the following maneuvers:

a. Spins.

b. Snap rolls or snap maneuvers.

BALLISTIC CONTROL SYSTEM LIMITATIONS.

To conserve catalyst bed life, continuous operation of the ballistic rocket motors should be limited to 3 seconds for ground or captive flight tests.

LANDING LIMITATIONS.

Information on sink speed - angle-of-attack limits for touchdown will be supplied when available.

CENTER-OF-GRAVITY LIMITATIONS.

Since the in-flight horizontal and vertical centers of gravity are automatically controlled as a result of the design of the engine and engine turbopump propellant systems, the horizontal CG normally is maintained within the forward and aft aerodynamic limits. The vertical CG will be within the engine adjustment provisions if all of the design equipment is installed. The engine and engine turbopump propellant jettison rate is automatically controlled to prevent an unsatisfactory CG condition. If airplane response during jettisoning indicates an unsatisfactory trim condition, selective jettisoning of the engine propellants should be accomplished.

WEIGHT LIMITATIONS.

The design of the airplane precludes the possibility of overloading at launch. The structural design weight for landing does not include engine propellants or engine turbopump monopropellant; therefore, they must be jettisoned before landing.

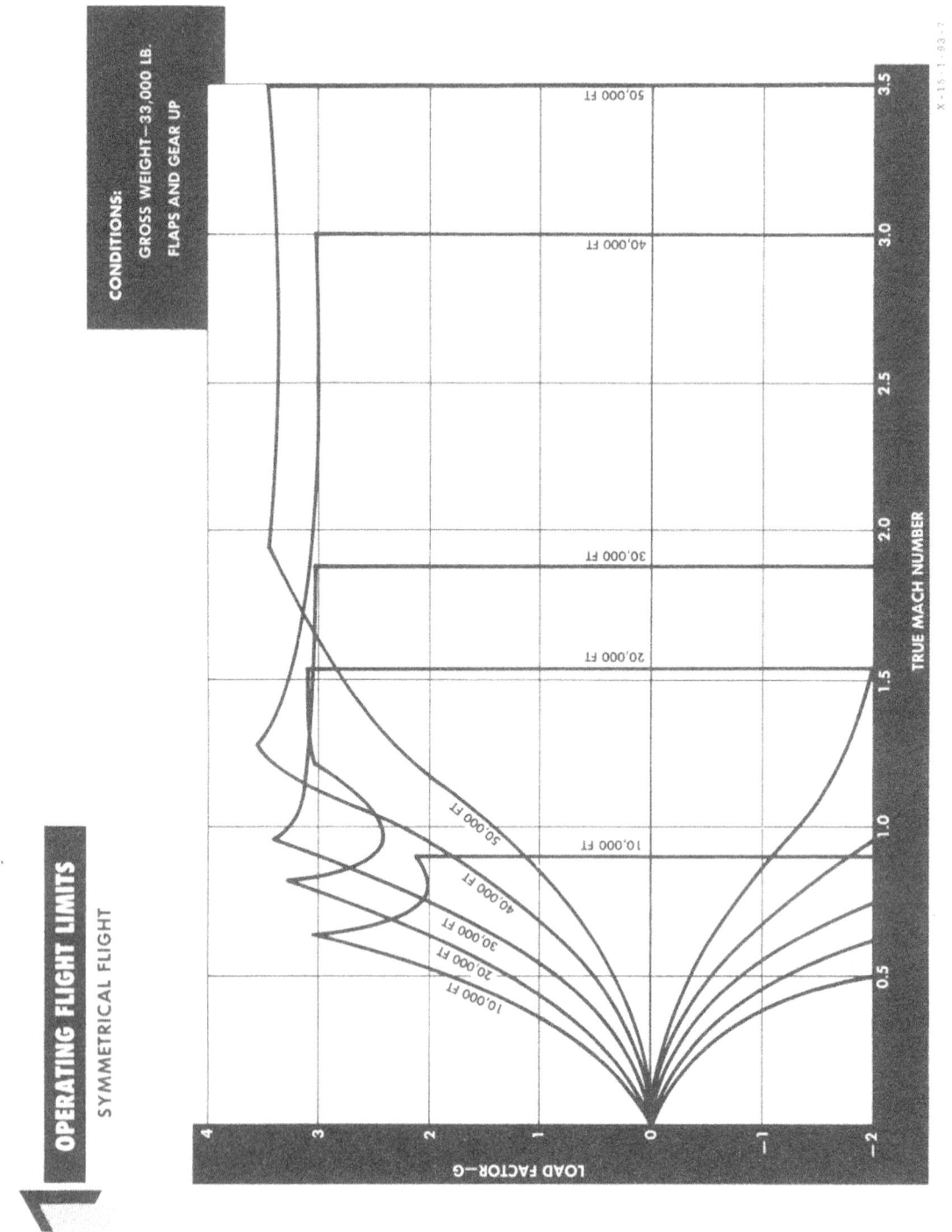

Figure 5-3. (Sheet 1 of 4)

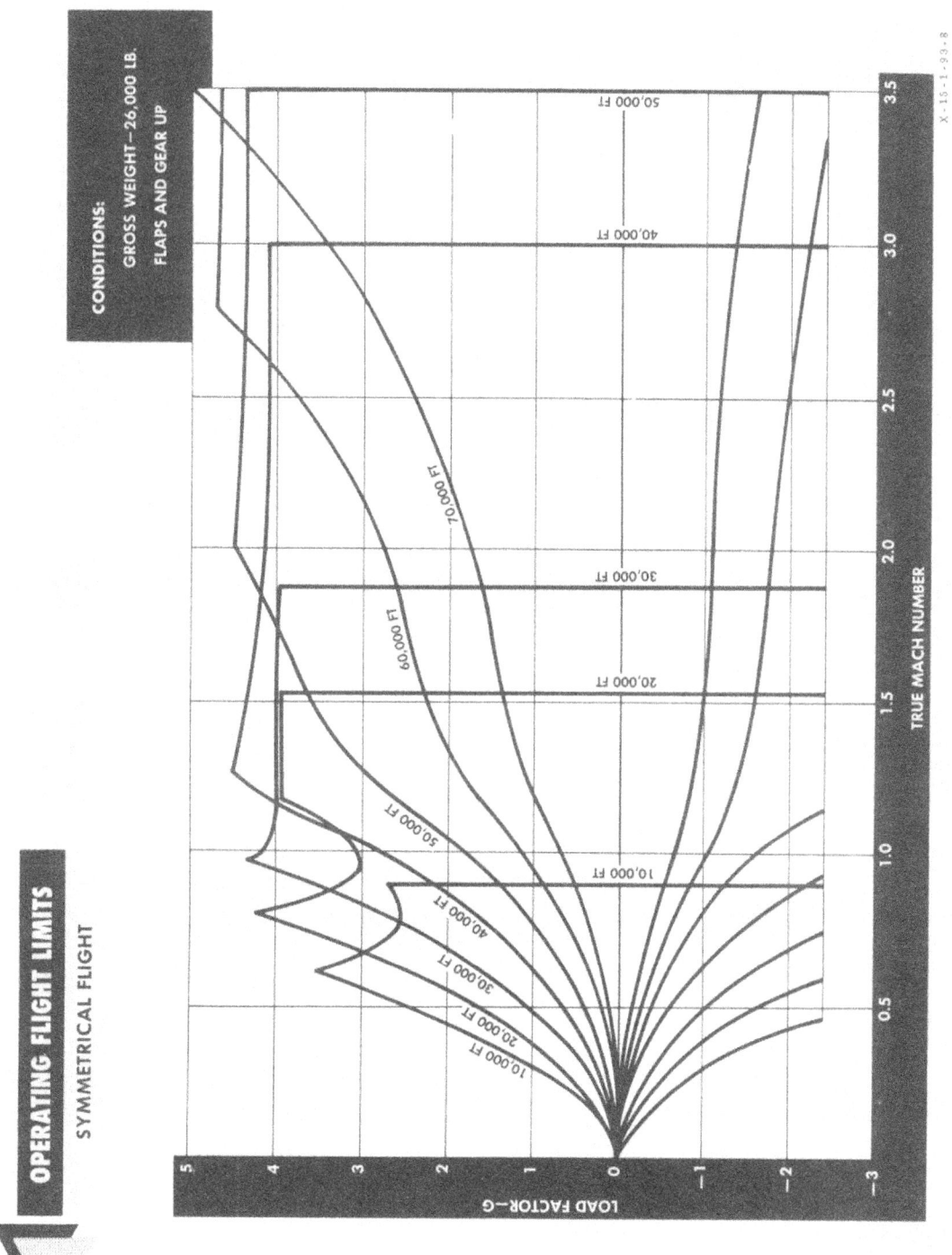

Figure 5-3. (Sheet 2 of 4)

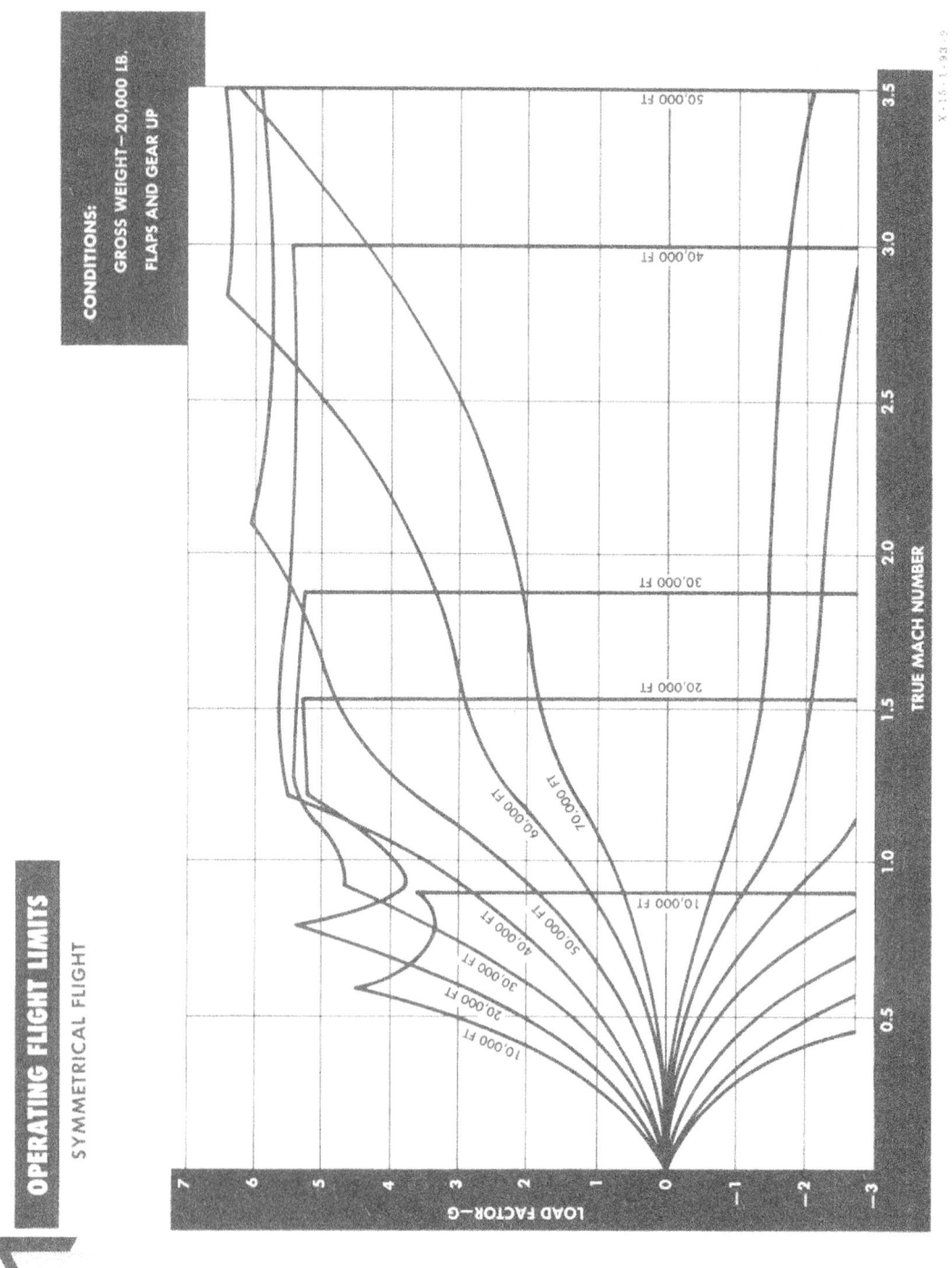

Figure 5-3. (Sheet 3 of 4)

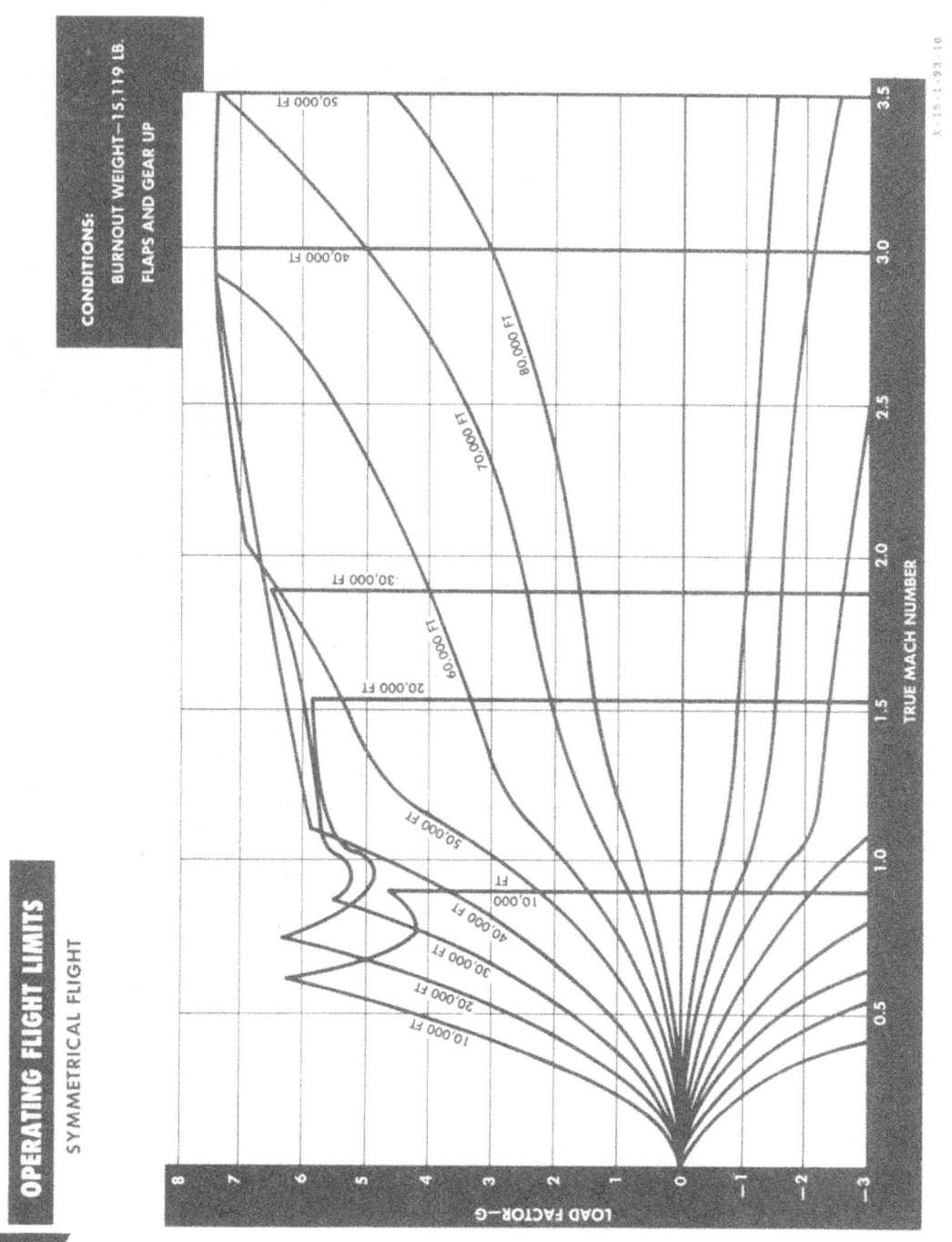

Figure 5-3. (Sheet 4 of 4)

Section V T.O. 1X-15-1

OPERATING FLIGHT LIMITS

ASYMMETRICAL FLIGHT

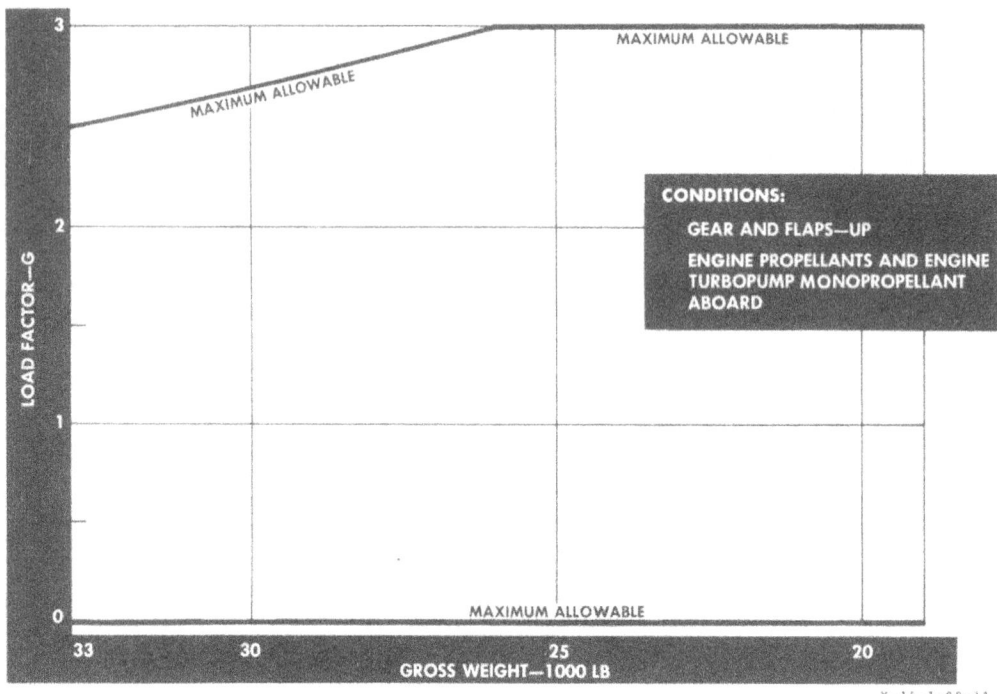

Figure 5-4.

This page intentionally left blank.

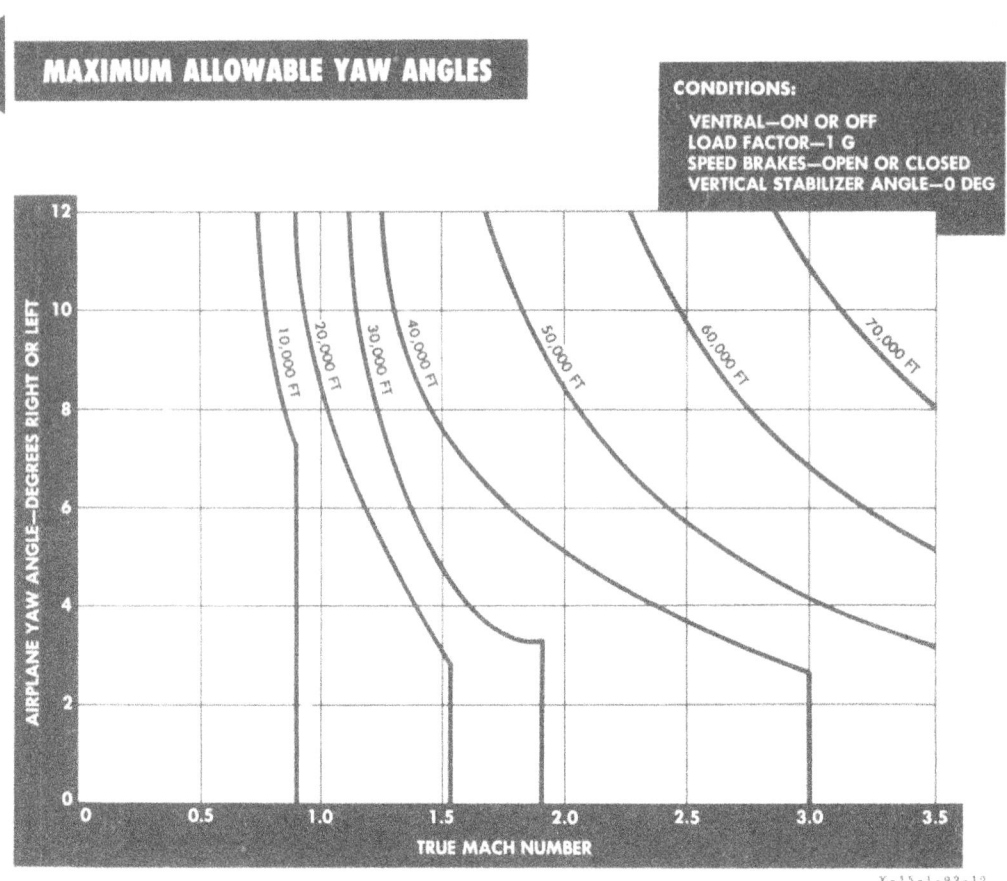

Figure 5-5.

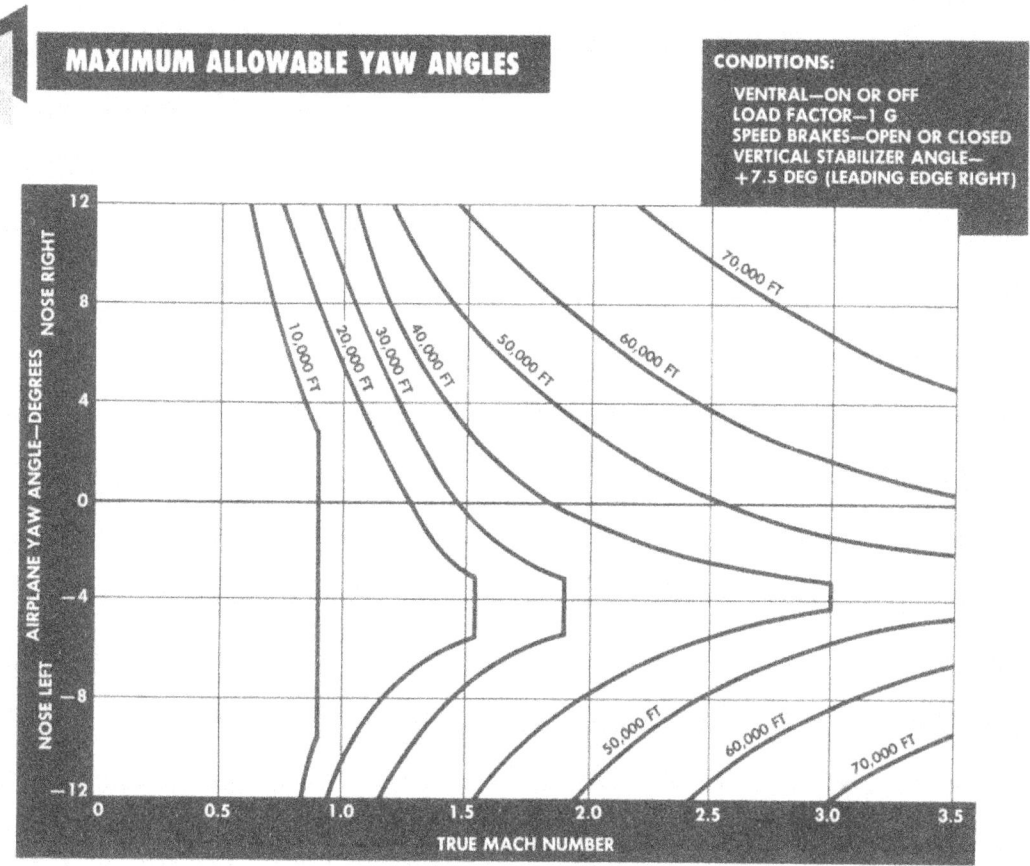

THIS PAGE INTENTIONALLY LEFT BLANK.

Flight Characteristics

SECTION VI

TABLE OF CONTENTS	PAGE		PAGE
Minimum Control Speeds	6-1	Flight Control Effectiveness	6-1
Spins and Spin Recovery	6-1	Flight Configuration Characteristics	6-3

MINIMUM CONTROL SPEEDS.

Minimum control speeds are not determined by a maximum lift coefficient, but by directional stability and estimated regions of buffet. Minimum control speeds for empty and launch weights are shown in figure 6-1.

SPINS AND SPIN RECOVERY.

WARNING

Intentional spins in this airplane are prohibited.

VENTRAL ATTACHED.

Fully developed spins are not apt to occur with the ventral attached, except possibly due to reduced directional stability at an angle of attack in excess of 20 degrees. If there is any yawing or rolling due to this reduced stability and a spin is entered, the stick should immediately be moved in the direction of spin rotation and full opposite rudder (against rotation) should be applied. When the spin has stopped, the stick should be centered and moved forward to prevent a spin in the opposite direction, and the rudder neutralized.

CAUTION

Do not immediately attempt to level the wings by moving the stick against the direction of rotation.

VENTRAL JETTISONED.

With the ventral jettisoned and the airplane at an angle of attack in excess of 17-1/2 degrees, a spin may be developed from which recovery may be difficult or impossible; therefore, jettisoning of this surface should be delayed as long as possible before entering the landing pattern. Recommended recovery is the same as when the ventral is attached: The stick should be moved in the direction of spin rotation and full opposite rudder should be applied. When rotation stops, the stick should be centered and moved forward to prevent a spin in the opposite direction, and the rudder neutralized.

FLIGHT CONTROL EFFECTIVENESS.

PITCH CONTROL.

For flight above approximately Mach 1.4, the maximum normal acceleration obtainable can be limited by the maximum stabilizer deflection available. Maximum normal acceleration can also be limited by an estimated initial buffet effect between Mach .5 and Mach 1.0. Below Mach .5, the limit is imposed by the reduced directional stability at angles of attack above 20 degrees. At low speeds, the angle of attack for buffet onset due to airflow separation on the wing is about 13 degrees. Turbulence behind the vertical stabilizer may cause considerable buffet at any angle of attack. Therefore, buffet onset should not be considered as a stall warning until buffet characteristics have been determined from flight experience.

ROLL CONTROL.

Adequate roll control is available throughout the Mach number range from high-speed flight to landing speeds. Rate of roll is limited to 50 degrees per second when the SAS roll gain selector knob is set on high gain. At lower angle of attack, there is slight favorable yaw (yaw in direction of roll application). As the angle of attack increases above Mach .6, there is a decrease in this yaw. At low Mach numbers and in the transonic region, there is very little change in roll effectiveness with angle of attack. Roll effectiveness increases with an increase in angle of attack above Mach 2.6. Roll coupling does not present a problem with or without SAS, except at extremely high Mach numbers at high altitudes. Roll coupling does not occur at as low an altitude with SAS in as it does with SAS out, and in either case, roll coupling would not occur below approximately 100,000 feet. Maximum rates of roll at other Mach numbers and altitudes are limited by the maximum differential horizontal stabilizer deflection.

MINIMUM CONTROL SPEEDS
GEAR AND FLAPS UP, LOWER VERTICAL STABILIZER ATTACHED

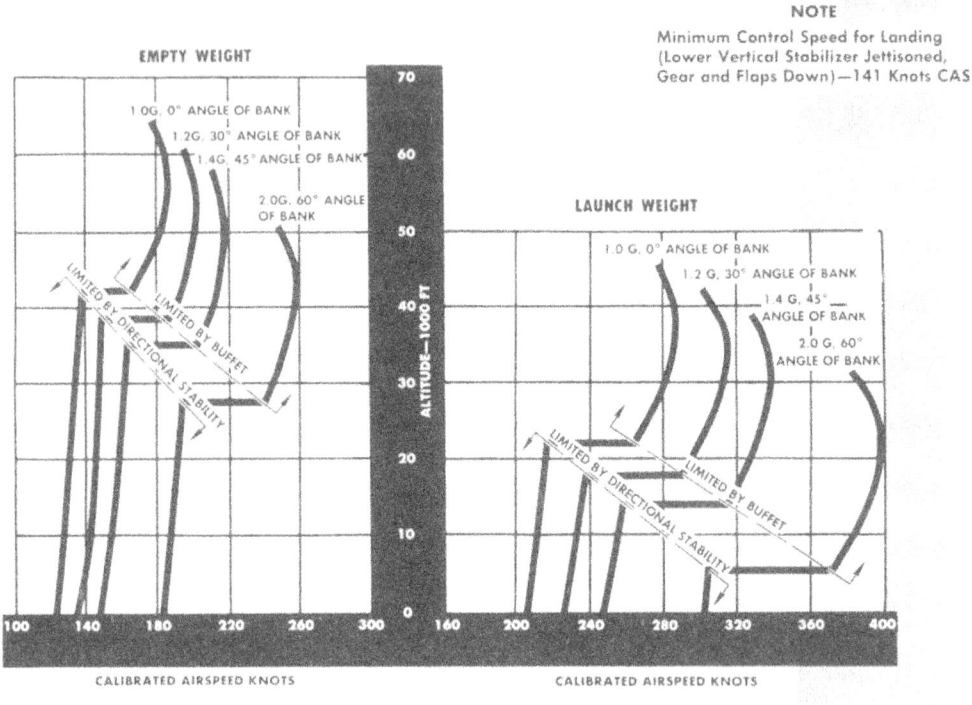

NOTE
Minimum Control Speed for Landing (Lower Vertical Stabilizer Jettisoned, Gear and Flaps Down)—141 Knots CAS

BASED ON: ESTIMATED DATA
DATA AS OF: 1 NOV 1958

Figure 6-1

YAW CONTROL.

Pilot feel is supplied as a function of surface deflection, and the variation of pedal force with pedal travel is linear. The upper and lower (ventral) vertical stabilizers give good directional control throughout the flight range. The ventral is effective in retaining stability at high angles of attack where the upper vertical stabilizer is partially blanketed by the wing and fuselage. At high Mach numbers and high angles of attack, this increase in effectiveness of the ventral produces an appreciable amount of roll in a direction opposite to the yaw. Although the ventral is jettisoned before landing, the upper vertical stabilizer supplies adequate directional control for low-speed flight up to 17-1/2 degrees angle of attack.

SPEED BRAKES.

At supersonic Mach numbers, in addition to increasing the effectiveness of the vertical stabilizers and improving directional stability, the speed brakes increase longitudinal stability. As angle of attack increases, the effect of the speed brakes on longitudinal and directional stability also increases. In addition to improving stability at high Mach numbers, the speed brakes obviously provide additional drag to the airplane. At subsonic Mach numbers, the speed brakes reduce both the longitudinal and the directional stability. During the glide, which would occur at lower altitudes, the basic drag of the airplane gives high rates of descent, and the increase in rate of descent due to use of the speed brakes is undesirable.

FLIGHT CONFIGURATION CHARACTERISTICS.

LAUNCH.

Launch from the carrier airplane can be accomplished satisfactorily for 1 G flight conditions during carrier airplane acceleration, climb, and cruise as well as the design launch conditions.

WARNING

- If a -5-degree stabilizer deflection is used for an empty-weight launch, the airplane will rotate about the forward part of the pylon, slide outboard, and hit the engine nacelles and wing tip. Full-weight launches with -5-degree stabilizer deflection are acceptable; however, it is recommended that a stabilizer deflection of zero to +5 degrees be used for all launches.

- The airplane rolls outboard upon release from the carrier. Roll control is very sensitive at this point, and no more than 2 degrees differential stabilizer deflection should be used if a correction is to be made.

GLIDE.

Figure 6-2 presents rate-of-descent - Mach number combinations for maximum glide distances at various altitudes and gross weights. A sample problem is provided with the chart.

RATE OF DESCENT AND MACH NUMBER FOR MAXIMUM

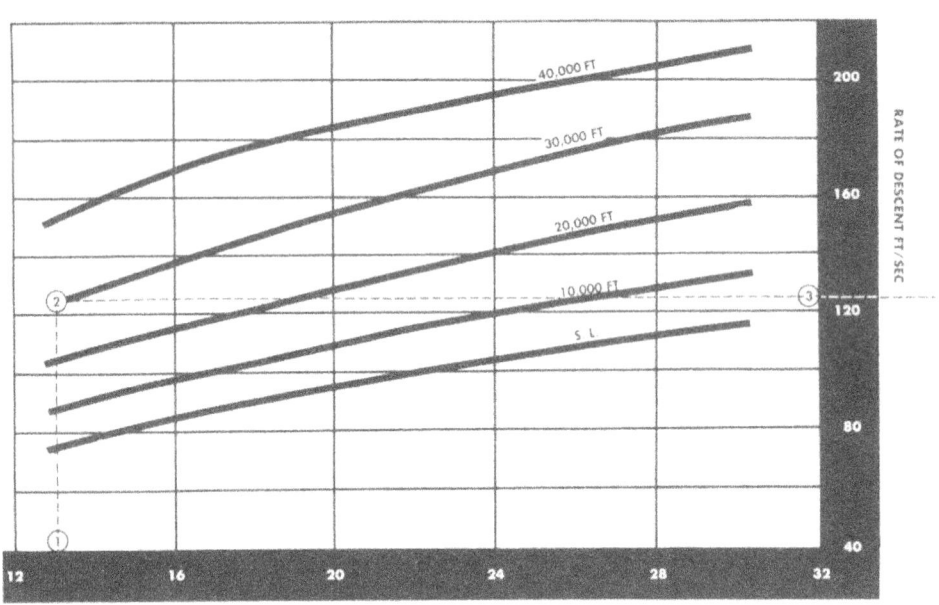

Figure 6-2

GLIDE DISTANCES

BASED ON: ESTIMATED DATA
DATA AS OF: 1 NOV. 1958

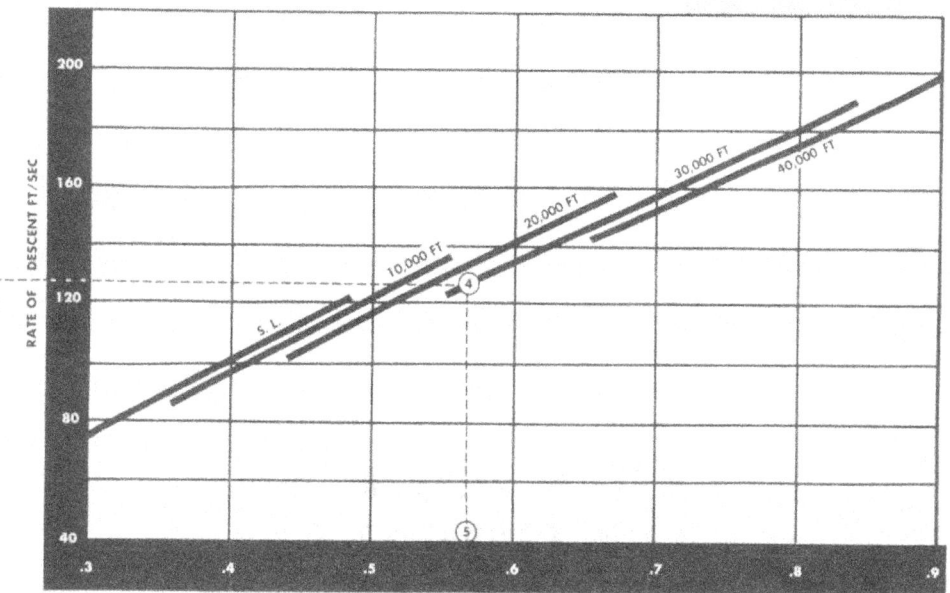

1. ENTER CHART FOR GROSS WEIGHT AT START OF GLIDE (FOR EXAMPLE, 13,000 LBS).
2. MOVE VERTICALLY UP TO CURVE REPRESENTING ALTITUDE AT START OF GLIDE (30,000 FEET).
3. MOVE HORIZONTALLY TO RIGHT TO RATE OF DESCENT LINE (125 FEET/SECOND).
4. CONTINUE HORIZONTALLY TO RIGHT TO ALTITUDE AT START OF GLIDE (30,000 FEET).
5. DROP VERTICALLY TO READ MACH NUMBER (.57) TO OBTAIN 125 FEET/SECOND RATE OF DESCENT.

X-15-1-93-4

THIS PAGE INTENTIONALLY LEFT BLANK.

NASA X-15 Ground Rescue

FLIGHT RESEARCH CENTER
NATIONAL AERONAUTICS and SPACE ADMINISTRATION

THIS PAGE INTENTIONALLY LEFT BLANK.

X-15 GROUND RESCUE

Compiled by Roger J. Barnicki
X-15 Rescue Training Officer

NASA Flight Research Center
Edwards, California

November 23, 1962

TABLE OF CONTENTS

	Page
INTRODUCTION	1
X-15 AIRPLANE	2
General Description	2
Propellant Supply System	5
Helium Pressurization and Pneumatic Control System	6
Hazards of Propellants and Gases Carried Aboard X-15	7
Hydrogen Peroxide	7
Anhydrous Ammonia	8
Liquid Oxygen	9
Liquid Nitrogen	10
Nitrogen and Helium Pressurizing Gases	10
Emergency Escape System	11
Ejection-Seat Systems	12
Canopy System	13
Operation of Canopy	14
Canopy-Jettison System	14
Parachute Assembly	15
Pilot's Emergency-Radio Beacon	15
X-15 A/P 22S-2 High Altitude, Full Pressure Suit Assembly	16
Description	16
Function	16
B-52 MODIFICATION	17
General Description	17
GROUND-RESCUE PROCEDURE	19
X-15 Pilot Removal	19
Securing X-15 Aircraft	21
X-15/B-52 Mated Condition	22
Removing and/or Rescuing the X-15 Pilot	22
Removing the B-52 Crew	22
X-15/B-52 EMERGENCY CONDITIONS AND PROCEDURES	23
Prior to Takeoff	23
In Flight	24
Landing	25
X-15 Emergency	26
FIGURES	27

INTRODUCTION

This manual provides pertinent information on ground-rescue procedure for the X-15 aircraft and the B-52A aircraft modified to carry the X-15. The information is presented to acquaint the fire-fighting and ground-rescue personnel with the X-15 and the modified B-52 configuration so that the X-15 pilot and the B-52 crew can be removed quickly and safely in an emergency.

X-15 AIRPLANE

GENERAL DESCRIPTION

The X-15 airplane is a single-place, mid-wing monoplane designed to supply fundamental research information on high aerodynamic-heating rates, stability, controllability, and physiological problems of high speeds and extreme altitudes. It is designed to be air-launched from a B-52 airplane. Distinguishing features of the X-15 are the slender cylindrical fuselage, with a short tapered wing located at the extreme rear; the cathedral horizontal stabilizer; and the wedge-shaped vertical stabilizers, above and below the fuselage (fig. 1). Inconel X is used for the skin of the fuselage, wings, and empennage.

A single YLR99-RM-1 rocket engine supplies the power for the X-15 airplane. Fuel used in this engine is a combination of anhydrous ammonia and liquid oxygen. The engine is mounted in the extreme aft section of the fuselage.

The X-15 airplane has a long, slender fuselage, tapering to a point at the forward end. The fuselage cross section is circular. A fairing is provided along each side of the fuselage to house the necessary plumbing, wiring, and controls. This eliminates the necessity of routing these items through the propellant tanks, which comprise the entire center portion of the fuselage (fig. 1).

The cockpit features a jettisonable, clamshell-type of canopy enclosure and a rocket-powered ejection seat. The windshield is an integral part of the canopy. It consists of a long, narrow window built into each side of the forward part of the canopy. Operation of the canopy is manual, with both spring and air assist bungees. A handle, inside the cockpit, below the right longeron, permits internal operation of the canopy (fig. 2). An external handle is provided in a well on the right side of the fuselage below the canopy longeron. Also within this well is an external emergency jettison handle, labeled FOR EMERGENCY USE ONLY, to forcibly jettison the canopy. An internal emergency jettison handle is provided on the right vertical instrument panel in the cockpit. The ballistic rocket-powered ejection seat features stabilization fins and ballistically powered stabilization booms to stabilize the seat in the event of ejection. An oxygen system is installed on the seat to provide breathing oxygen under normal conditions. During the ejection cycle, the seat oxygen system also provides suit pressurization. At the point of seat-pilot separation, a self-contained suit-oxygen system provides the pilot with breathing oxygen for the remainder of the ejection descent.

The X-15 landing gear is a tricycle configuration. The nose gear is located well forward on the fuselage, and the main gear is located far to the rear on either side of the fuselage. The nose gear is nonsteerable and includes dual wheels and high-pressure tubeless tires. The wheels are splined to a common axle shaft which rotates on tapered bearings in the landing-gear shock strut. Each main landing gear consists of a short, broad ski-type skid, attached at the end of an inflexible steel strut. Cushioning action is provided by an air-oil cylinder at the top of the strut. The landing gear is manually retracted and is locked in the retracted position by the ground crew before flight. It is mechanically released by the pilot, and is actuated down by gravity and aerodynamic forces. Once extended, the landing gear is locked into position and cannot be retracted by the pilot.

The X-15 has a short, tapered, low-aspect-ratio, thin wing. Notable features are the blunt leading and trailing edges, the squared-off wing tips, and the absence of ailerons. A wing flap forms the inboard trailing edge of each wing panel. The wing is mounted far to the rear along the fuselage horizontal centerline (fig. 1).

The empennage consists of an all-movable horizontal stabilizer and a two-piece vertical stabilizer, both above and below the fuselage. The sweptback horizontal stabilizer has a 15° cathedral. It deflects as a unit for pitch control and differentially for roll control. The movable (outboard) portions of both the upper and lower vertical stabilizers provide yaw control. The movable portion of the lower vertical stabilizer is jettisoned before landing. The fixed (inboard) portion of the vertical stabilizers, both above and below the fuselage, splits at the aft end to form hydraulically operated speed brakes. The horizontal stabilizer and the movable vertical stabilizers are actuated by irreversible, dual hydraulic actuators for control within the atmosphere.

During flight at extreme altitudes, where aerodynamic controls are ineffective, a system of ballistic control rockets is provided. Eight controllable, monopropellant rockets, mounted in pairs in the extreme nose of the airplane, provide control for pitch and yaw. A pair of rockets is mounted in each wing to control roll.

The X-15 fuselage is a semimonocoque design. It is a combination of welded and riveted construction. The cylindrical shape of the fuselage provides the maximum strength, together with the best utilization of the internal area. The forward part of the fuselage contains the nosewheel well and the cockpit and equipment section. The entire intermediate section of the fuselage consists of the liquid-oxygen and fuel tanks. The rear part of the fuselage contains the engine and associated equipment. The rear section also contains the hydraulic actuators, valves, and associated equipment for operation of the vertical and horizontal stabilizers.

The extreme forward portion of the X-15 nose, containing part of the airspeed system, is removable. This provides access to airspeed equipment, as well as interchangeable nose sections. The flow-direction sensor (ball nose) is used to gain information on stability and control of the airplane and is required for high-speed runs. Immediately behind this section are eight monopropellant attitude rockets (fig. 3), arranged in pairs. These ballistic-control rockets provide yaw and pitch control at altitudes where the aerodynamic controls become ineffective. They use hydrogen peroxide as fuel. The high-altitude data compartment, containing the high-altitude recorder, is located between the ballistic control units and the nosewheel well. Insulation and radiation blankets provide protection for the compartment and nosewheel well. The nose landing gear occupies the remaining space rearward to the cockpit forward bulkhead.

The cockpit and equipment section is of an inner and outer concentric-shell construction. The outer shell forms the fuselage skin. The inner shell is the pressure barrier for the cockpit and pressurized equipment sections. The outer shell is of Inconel X; the inner pressure shell is of aluminum. Between these shells is an insulation blanket and radiation shield which insulate the cockpit and equipment section against aerodynamic heating.

An electronic equipment elevator (fig. 3) is installed in the equipment bay aft of the cockpit. The elevator consists of three shelves mounted in a frame. It is lowered into place, guided by rails attached to the fuselage structure.

Immediately to the rear of the electronic-equipment section (fig. 3) is a compartment containing the two auxiliary power units (APU), several small tanks, and related equipment. The auxiliary power units supply power for the airplane dual hydraulic and electrical systems.

To the rear of the cockpit and equipment section are the propellant tanks (figs. 3 and 4). These tanks form the fuselage structure in this area. The tanks are of welded construction, having torus frames and bulkheads to provide strength. A liquid-oxygen tank forms the forward part of this section. Installed through the center of the liquid-oxygen tank is a cylindrical bottle containing helium (fig. 4), which is used for pressurizing and purging the propellant tanks. The area immediately aft of the liquid-oxygen tank contains a small equipment compartment and a helium tank used for engine purging and engine-pump hydrogen-peroxide pressurization.

Behind this compartment, and forming the balance of the intermediate part of the fuselage, is the fuel tank (figs. 3 and 4). The lugs for the wing-attachment fittings are welded into this portion of the fuselage. The rear section of the fuselage contains the YLR99-RM-1 engine and related equipment. A fire wall separates the engine compartment from the propellant tanks. The fuselage tail cone is removable, providing access for removal or installation of the engine.

PROPELLANT SUPPLY SYSTEM

The YLR99 engine propellants are anhydrous ammonia and liquid oxygen. Hydrogen peroxide is also used as a monopropellant to drive engine accessories and fill other power requirements.

The fuel oxidizer is liquid oxygen, which is stored in a 1,000-gallon cylindrical container as an integral part of the fuselage. The helium tank is in the center of the liquid-oxygen tank (fig. 4). The helium-pressurization system pressurizes the liquid-oxygen tank to 46 to 53 psig, providing for expulsion of the liquid oxygen to the engine or out through the jettison lines.

The liquid-oxygen tank is divided into three interconnected compartments, check-valved to permit venting forward at the top and oxygen flowing rearward at the bottom. The check valves at the bottom of the compartments insure that all forward transfer occurs at the compartment top, which, in turn, insures complete liquid filling. These valves also prevent rapid forward transfer in case of decelerative maneuvers, thus preventing an uncontrollable center-of-gravity shift.

Two pneumatically operated valves, a vent pressure-relief valve and a tank pressure-regulating and pressurization valve, control the flow of the pressurizing gas (helium) to the liquid-oxygen tank. When the cockpit control handle of the pressurization-system selector valve is in VENT position, the normally closed vent pressure-relief valve is pneumatically operated to OPEN, and the normally open pressure-regulating and pressurization valve is pneumatically operated to CLOSED. When the cockpit control handle is placed at PRESSURIZE, control-gas pressure is shut off from the vent pressure-relief valve and the pressure-regulating and pressurization valve. This places the vent valve in a closed, nonvent condition, except for (relieving) excess pressure, and the pressure-regulating valve in an open, pressurizing condition. Helium at 48 psi enters the liquid-oxygen tank and expels liquid oxygen through the feed line to the engine. When the cockpit control handle is placed at JETTISON, the liquid-oxygen pneumatically actuated jettison valve is actuated open, and pressurized liquid oxygen is forced overboard through the jettison lines.

The rocket-propellant fuel, liquid anhydrous ammonia (NH_3), is stored in a 1,447-gallon cylindrical tank as an integral part of the fuselage (fig. 4). The tank is pressurized to 46 to 53 psig by the helium pressure system to provide for the expulsion of fuel to the engine or, if necessary, out through the jettison lines.

The ammonia tank is divided into three compartments. The compartment-filling sequence is forward, center, then rearward, which is the direct opposite of the filling sequence of the liquid-oxygen tank. The tank vent check valves permit venting from front to rear. The transfer tubes allow fuel flow from rear to front for fuel transfer to the engine or, if necessary, to jettison. This causes the propellant center of gravity to converge on the airplane center of gravity and permits satisfactory balance during either horizontal or vertical flight. Ammonia tank pressurization, vent, and jettison are controlled by valves in an identical manner to that of the liquid-oxygen tank.

Hydrogen peroxide (H_2O_2), a monopropellant, supplies the necessary steam through a decomposition process to power the turbine that drives the engine fuel and oxidizer pumps. The hydrogen peroxide is stored in a 10-cubic-foot spherical tank aft of the fuel tank (fig. 4). The H_2O_2 tank is pressurized to 560 to 615 psi by a helium pressure system separate from the propellant-tank system. This pressure expels the H_2O_2 to the gas generator where it is decomposed by a catalyst, producing steam for the turbine or (when necessary) expelling the H_2O_2 overboard through the jettison line.

A pneumatically operated pressurization vent valve controls the pressurization of the H_2O_2 tank. When the cockpit control handle of the pressurization-system selector valve is in VENT position, the normally closed vent-valve section opens and the normally open pressurizing valve section closes. This vents the tank and prevents the system from becoming pressurized. When the control handle is moved to PRESSURIZE, gas pressure is removed from the control valves and the valves are spring-forced to the normally closed position in the vent section. When the control handle is moved to JETTISON, the hydrogen-peroxide jettison valve is actuated open and the pressurized H_2O_2 is forced overboard through the jettison line.

Helium Pressurization and Pneumatic Control System

Helium is used to pressurize the liquid-oxygen, ammonia, and hydrogen-peroxide tanks. It also provides pressure for all pneumatic valves in the fuel system and the pneumatic engine control valves; supplies engine purge gas; and, in case of fire in the engine compartment, is used to make the atmosphere of the engine bay inert.

The helium gas is stored in eight tanks (fig. 4). Helium to pressurize the fuel and oxidizer tanks is stored in a cylindrical tank (FOX source), positioned in the center of the liquid-oxygen tank, and is pressurized to 3,600 psig. Helium to pressurize the hydrogen-peroxide tank and to supply engine control gas, engine purge gas, and pneumatic pressure where needed is contained in four spherical tanks. One tank is located between the liquid-oxygen and ammonia tanks, one is immediately aft of the ammonia tank, and two are aft of the hydrogen-peroxide tank on either side of the engine compartment. All four of these source tanks are pressurized to 3,600 psig. The three remaining tanks are located just forward of the liquid-oxygen tank and below the aircraft auxiliary power units. One of these tanks provides pressurizing gas for the 26-gallon liquid-nitrogen tank used for supplying cooling and pressurization to the forward compartments. This helium source is pressurized to 3,600 psig and delivers regulated helium gas to the liquid-nitrogen tank at 67 psig. The two remaining tanks are for pressurizing the two 13-gallon hydrogen-peroxide tanks of the auxiliary power units and ballistic control systems. These two helium tanks are pressurized to 3,900 psig and deliver regulated pressure at 575 psig to the two hydrogen-peroxide tanks for operation of the auxiliary power units and ballistic control rockets.

Hazards of Propellants and Gases Carried Aboard X-15

Hydrogen peroxide.- Hydrogen peroxide is a clear, water-like liquid which has a faint bluish cast when seen in a large quantity, and a sharp, acrid odor. Concentrated hydrogen peroxide, if handled properly, is considered to be no more dangerous than most other corrosive and flammable fuels. However, it is considered a fire hazard, and contact of the solution with any combustible material, including most fabrics, usually results in a spontaneously ignited fire. The fire cannot be smothered because of the oxygen supplied during decomposition. Therefore, water, which dilutes the peroxide and cools the flame, is the only satisfactory fire-extinguishing agent. Other agents, such as caustic and foam, may combine with hydrogen peroxide to form dangerous mixtures.

Normally, hydrogen peroxide is not explosive; however, it has been known to detonate if subjected to a violent shock load or heated to high temperature while confined. Also, vapors over concentrated solutions have been exploded with hot wire.

Concentrated hydrogen peroxide, although not toxic, is a strong oxidizing agent and inorganic irritant. Contact with the skin should be avoided. Its vapor causes irritation to the skin, mucuous membrane, and the eyes. If contact with the skin is relatively short, no injury will result beyond a temporary bleaching accompanied by a tingling or burning sensation. Contact for longer periods results in redness and, possibly, blistering. No permanent injury has been observed from exposure of the human skin to hydrogen peroxide.

Inhalation of hydrogen-peroxide vapor may cause extreme irritation and inflammation in the nose and throat and, if prolonged, can produce swelling or accumulation of fluid in the lungs. Exposure for short periods has not caused lasting harm.

Exposure of the eyes to splashes from concentrated hydrogen peroxide may cause injury such as ulceration of the cornea, with resulting blindness. It should be noted particularly that injury to the eyes may not be apparent immediately. Corneal ulceration may appear even a week or more after exposure. It must be realized, however, that vapor will sting slightly and cause watering of the eye, but no real damage.

FIRST AID The primary consideration in first aid in hydrogen-peroxide contamination is prompt removal of the patient from any area where there is a high concentration of vapor or mist in the air, and removal from the skin of any solution which may be present. If the solution has contacted the skin, it should be washed off as quickly as possible with large quantities of water. Burns require treatment by a physician. If the solution has entered the eyes, they should be irrigated immediately and freely with water for at least 15 minutes. The patient should be referred promptly to a physician, preferably an eye specialist. If the gas is inhaled, the patient should be removed at once from further exposure. If irritation of nose and throat is severe, a physician should be consulted.

Anhydrous ammonia.- Anhydrous (waterless) ammonia (NH_3) is formed by the combination of the two gaseous elements, nitrogen and hydrogen. Liquid ammonia is colorless; it boils at a temperature of -28° F at atmospheric pressure, and freezes to a white crystalline mass of -107° F. Gaseous ammonia is also colorless and has an extremely sharp, penetrating, pungent, and unpleasant odor.

Concentrated ammonia is both toxic and flammable and must be considered extremely dangerous. It is imperative that all persons handling ammonia be thoroughly familiar with the hazards involved. Although ammonia is sometimes classified as a nonflammable compressed gas, the gas is flammable in air in comparatively high concentrations (16 to 25 percent by volume). Since ammonia has a penetrating, suffocating odor, and is strongly irritating, it is not likely that a person will voluntarily remain in a contaminated atmosphere. Irritation to the eyes, respiratory tract, and throat results from concentrations as low as 0.05 to 0.1 percent. A concentration of 0.2 percent produces convulsive coughing and may be fatal after an exposure of less than one-half hour. Ammonia gas in a concentration of 0.6 to 1.0 percent of volume is lethal within a few minutes. The maximum concentration tolerated by the skin for more than a few seconds is 2 percent. Concentrations on the order of 50 parts per million (0.005 percent) can be detected by the average person.

Ammonia, either as an anhydrous liquid or gas, is a strong irritant, capable of inflicting severe irritation to the surface tissues of the eyes, nose, throat, and lungs. One of the most hazardous effects of ammonia is its effect on the eyes. When the eye is sufficiently exposed to ammonia, the iris becomes congested and hemmorhaging begins, often resulting in permanent blindness. In addition to its caustic action, liquid ammonia may cause severe injury to the skin by freezing the tissue. If handled improperly, liquid ammonia can cause irreparable damage, or death.

Canister-type industrial gas masks with full face pieces are adequate protection where ammonia concentration does not exceed 3 percent. For emergencies, or where concentrations exceed 3 percent, a closed-system supplied-air respirator with a full face piece should be used.

FIRST AID Of paramount importance in the administration of first aid in cases of anhydrous-ammonia contamination is speed in removing the ammonia from contact with the patient and in moving the patient to an uncontaminated atmosphere. Artificial respiration must be applied immediately to a person who has been overcome by ammonia, using only the Shaefer method. The use of Pulmotor apparatus or any manipulation of the arms should be avoided to prevent further irritation to the lungs. To prevent severe lung congestion, pure oxygen should be administered to the patient as soon as possible. If exposure has been mild, relief may be had by stuffing 2-percent boric-acid solution up the nose and rinsing the mouth thoroughly. If swallowing is possible, the patient should be encouraged to drink large quantities of 1/2-percent citric-acid solution or lemonade. Giving anything orally to an unconscious person should never be attempted.

All contaminated clothing should be removed from the patient at once and the affected areas washed with large quantities of cool water. Under no conditions should salves or ointments be applied to skin or mucous-membrane burns for 24 hours following the injury.

If ammonia has contacted the eyes, irrigation should be started immediately with copious quantities of cool, clean water, irrigating continuously for 15 minutes, waiting 10 minutes, and irrigating again for 5 minutes. This procedure should be continued for 1 hour. After the ammonia has been removed as thoroughly as possible, the patient should be kept warm and quiet. Inadequately treated cases of severe exposure can occasionally develop into a secondary bronchopneumonia, which could be fatal. <u>In all cases</u>, a doctor should be notified at once.

<u>Liquid oxygen</u>.- Liquid oxygen is very pale blue and somewhat less viscous than water. Since the boiling point of liquid oxygen is -297° F at standard atmospheric conditions, it is constantly evaporating into gaseous oxygen which is colorless and odorless. Pure oxygen, either in a liquid or gaseous state, does not burn or explode. However, liquid oxygen, when mixed with organic liquids, creates a very sensitive high explosive. Gaseous oxygen will support combustion in a violent manner. Liquid oxygen is extremely dangerous when spilled in a confined area; the gaseous oxygen produced may remain in relatively high concentration while flowing considerable distances. If an oxygen-fed fire occurs, the only positive way of stopping it is by shutting off the oxygen supply. If this is impossible, large quantities of water may cool the flame enough to extinguish it. Liquid oxygen mixed with grease, oils, petroleum-derivative fuels, or alcohol, forms a highly impact-sensitive gel which is about equivalent to nitroglycerin in explosive energy. This gel, when formed, has the appearance of a dirty white slush and may be detonated by a spark or flame as well as a mechanical shock. It is, therefore, imperative that a gel, if accidentally formed, should not be disturbed and the area should be evacuated until the liquid oxygen evaporates.

The only physiological hazard to be considered in handling liquid oxygen is the effect of the extreme cold, which destroys living tissue and produces an injury similar to a burn. In addition, freezing can restrict blood circulation and cause gangrene, especially in the extremities. Sufficient liquid oxygen splashed into the eyes will cause blindness. Gaseous oxygen is not toxic, and inhalation has an exhilarating effect.

FIRST AID In all cases of liquid-oxygen contamination, the patient should be removed from the contaminated atmosphere as quickly as possible. If liquid oxygen should contact the body, the affected area should be immediately flushed with large quantities of water. This action will serve to wash off the liquid oxygen and also warm the skin.

If the eyes have been contaminated with liquid oxygen, they should be irrigated with a 3-percent boric-acid solution. Competent medical attention is required as soon after first-aid treatment as possible. Ordinary clothing becomes easily saturated with gaseous oxygen and should be aired for a minimum of 2 hours and, preferably, for as long as 12 hours. Any person who has been exposed to concentrations of gaseous oxygen should not smoke or approach any source of ignition for 2 hours, unless a complete change of clothing has been made.

Liquid nitrogen.- Liquid nitrogen is clear in color and somewhat less viscous than water. Since the boiling point of liquid nitrogen is -320° F at standard atmospheric conditions, it is constantly evaporating into gaseous nitrogen which is colorless and odorless.

Nitrogen, either in a liquid or gaseous state, is chemically inactive, and presents no hazards except those resulting from its low temperature and exclusion of oxygen in the immediate area. The extreme coldness of liquid nitrogen will destroy living tissue and produce an injury similar to a burn. In addition, freezing can restrict blood circulation and cause gangrene, especially in the extremities. Sufficient liquid nitrogen splashed into the eyes will cause blindness. Liquid nitrogen, although not toxic in the usual sense, can cause asphyxiation by exclusion of oxygen from the immediate environment.

FIRST AID In all cases of liquid-nitrogen contamination, the patient should be removed from the contaminated atmosphere as quickly as possible. Artificial respiration must be applied immediately to a person who has been overcome by a nitrogen-rich atmosphere.

If liquid nitrogen should contact the body, the affected area should be immediately flushed with large quantities of water. This action washes off the liquid nitrogen and also warms the skin.

If the eyes have been contaminated, they should be irrigated with a 3-percent boric-acid solution.

Competent medical attention is required as soon as possible after first-aid treatment.

Nitrogen and helium pressurizing gases.- Nitrogen (N_2) and helium (He) are both odorless, colorless gases used for pressurizing rocket-propulsion systems. Helium is chemically inert and nitrogen is chemically inactive. Neither gas is corrosive, explosive, nor combustible.

The only possible hazards arising from the use of either nitrogen or helium stem from (1) high storage and transfer pressures normally employed, or (2) shortage of breathing oxygen because of extremely high concentrations of either gas.

EMERGENCY ESCAPE SYSTEM

The emergency escape system quickly and safely separates the pilot from the airplane in an emergency which would require bailout. The system includes an integrated jettisonable canopy and ejection seat controlled by the ejection handle assemblies on the pilot's ejection seat. This combined system consists of the canopy system, emergency canopy-jettison system, manually operated control system, and gas-pressure initiating system. The ejection-seat system contains the ejection seat, a ballistic-rocket type of catapult, pilot restraints, the restraint-release system, the parachute assembly, and a seat-mounted oxygen system.

Escape from the aircraft is accomplished by a single movement of the ejection handles on the ejection seat. A sequencing system delays firing of the ejection-seat ballistic-rocket catapult until the cockpit canopy has cleared the airplane. This sequencing system also permits the cockpit canopy to be jettisoned without firing the seat rocket catapult, provided the ejection handles have not been raised. Injury to the pilot from post-ejection accelerations and wind blast forces is prevented by the restraint system. The ejection handles and armrests provide protection and restraint for the pilot's hands and arms; the integrated harness and seat cushion provide restraint for the pilot's body; and the foot-restraint system and leg guards provide restraint for the pilot's feet and legs. The restraint devices are brought into position by pilot arm and leg motions prior to ejection.

The ejection seat is stabilized after it leaves the airplane by the seat-stabilization system, consisting of the stabilization fins and stabilization booms (fig. 5). These are aerodynamic devices that prevent the seat from tumbling or spinning during free fall after ejection. The stabilization devices are extended pneumatically and are triggered by the motion of the seat as it moves upward on the ejection rails.

The pilot is released from the ejection seat during free fall by an altitude-sensing device which energizes the restraint-release system and deploys the pilot's parachute. If ejection occurs at an altitude less than 15,000 feet, the same device functions 3 seconds after the ejection seat leaves the airplane. A manual emergency restraint-release control is also provided on the right-hand forward corner of the seat (figs. 5 and 6).

The parachute system consists of a flat circular parachute, packed in a Fiberglas container, and a jettisonable headrest which forcibly deploys the pilot's chute and assists in extracting it from the container. The escape controls are arranged so that actuation of the manual emergency restraint-release control while the seat is in the airplane causes the headrest to be jettisoned without deploying the pilot's parachute.

EJECTION-SEAT SYSTEMS

Manual release of the restraint system by ground-rescue personnel is accomplished in the following manner:

Release integrated shoulder-harness parachute-riser assembly and parachute "D" ring retainer quick disconnects (fig. 7(a)) by squeezing the latches on both sides of the locking assembly and sliding the assembly away from the harness attaching rings. Remove the lap-belt buckle of the integrated parachute-harness seat belt (fig. 7(b)) by grasping the buckle on both outboard sides and pulling up and away from the pilot. The buckle assembly must be removed from the lap belt before the pilot is free of the seat.

Depress foot-manacle release buttons (fig. 7(c)) if pilot's feet are locked in foot restraints. (One release button is located on each forward edge of the ejection seat.) Slide foot forward as each button is depressed. Disconnect or actuate pilot's emergency-oxygen system, depending upon the concentration of ammonia vapors. To disconnect the emergency-oxygen actuating cable, located on the right side of the back-pan assembly and connected to an I-bolt by a pip-pin assembly, depress the pip-pin release button at the ejection-handle trunnion joint, and remove from the I-bolt. If the pip pin is not disconnected, the emergency-oxygen supply will be actuated as the pilot raises from the seat. Open helmet visor if emergency-oxygen system is not armed. Shut off the aircraft oxygen system by moving the rotating oxygen selector, located at the forward left-side panel of the seat, to the off position. Disconnect personal-lead quick disconnect (fig. 7(d)), located on the inner left side of the seat, by actuating the release with a squeezing, upward motion. Install ejection handle and restraint emergency-release-handle safety pins (fig. 5). Pilot is then ready for exit or removal from the cockpit.

The emergency restraint-release system is to be used only when time does not permit safe pilot exit in the aforementioned manner. The emergency restraint system is actuated by pulling the restraint emergency-release handle, located on the right forward end of the seat frame (figs. 5 and 6). Pulling the restraint emergency-release handle causes:

Disengagement of headrest from parachute.
Release of foot restraints.
Unhooking of pilot's integrated harness at lap-belt fittings
   located on each side of seat.

Release of personal-lead quick disconnect.

Release of shoulder harness as the headrest is jettisoned upward, which allows the harness to fall free when harness retainer pins attached to the seat headrest are pulled out.

The parachute remains with the pilot when released in this manner.

WARNING The restraint emergency-release system must not be used until the canopy has been fully removed from the aircraft, either by the canopy-jettison system or by manually pushing the canopy off the aircraft. Do not stand above seat when the restraint is pulled; the headrest is ejected straight up with explosive force.

Install the ejection-handle safety pins or cut the ejection-seat initiator hose to disarm the seat when the canopy is to be removed from the aircraft. The access panel to the initiator hose is located on the top of the fuselage, just aft of the canopy. The hose "cut" point is painted red.

After the specified procedures have been accomplished, remove pilot from cockpit.

CANOPY SYSTEM

The canopy system consists of two interconnected subsystems: a mechanical system, and a jettison system. The mechanical system is manually controlled and operated by linkages, pneumatic balance springs, and bungees. This system is used for normal opening and closing of the canopy. The jettison system is manually controlled and pneumatically operated by explosive charges. It is used for removing the canopy from the airplane in an emergency. This system consists of an exactor, a canopy remover, two M-3A1 initiators, two M-5 initiators, a thrust-augmentation unit, and a thruster-sequencing valve.

The canopy has a compound motion for actuation, that is, a short linear motion to disengage the mechanical locks on the left- and right-cockpit longerons, and an upward or downward rotary motion for opening or closing the canopy.

Normal operation of the canopy system is accomplished manually, partially through a mechanical linkage, and partially through direct physical contact. Operation of the canopy from within the cockpit is accomplished by using the canopy handle on the right forward side of the cockpit under the longeron (fig. 2). The canopy may be opened by rotating the canopy handle inboard (clockwise) until the handle reaches the end of its travel, then pulling it toward the rear. This will slide the canopy aft 1 inch and position it for the second phase of the operation.

Operation of Canopy

The second phase of the operation, that of rotation, is essentially a two-position cycle. The canopy has a pneumatic counterbalance (pneumatic balance springs and damper bungee) such that a balance point exists with the canopy about 10° to 15° open. Therefore, when opening the canopy, force is required only for the first 10° to 15° of rotation, after which, canopy rotation is automatic.

The canopy may be operated externally through the use of the canopy handle on the right side of the airplane, below the longeron near the leading edge of the canopy (figs. 2 and 6). The handle is reached by pushing the circular button on the door-latch mechanism. The canopy can then be opened by pulling the handle outboard, releasing it from its clip and stop, and rotating the handle upward to move the canopy to the rear 1 inch. The canopy must then be raised by being grasped at the leading edge and manually moved to its open position.

Canopy-Jettison System

The canopy-jettison system is manually controlled by the seat-ejection handles and the internal and external CANOPY EMER JETT handles. Raising the ejection handles on the pilot's ejection seat fires the canopy-remover initiator on the canopy deck through a linkage arrangement which, in turn, fires the canopy remover. This unlocks the canopy and moves it aft 1 inch. As the canopy moves aft, the canopy thruster-sequencing valve is actuated by the canopy, allowing gas pressure from the canopy initiator to pass through the sequencing valve and fire the initiator adjacent to the valve. Gas pressure from this initiator is directed forward to fire the booster initiator adjacent to the canopy auxiliary thruster. Gas pressure from the booster initiator is then directed to the canopy thruster cylinder, forcing the piston to extend. As the piston extends, the forward end of the canopy is forced upward. The canopy remover then strikes the canopy impact fitting, forcing the canopy upward and free of the airplane (figs. 8(a) and 8(b)). All of these movements are simultaneous, with the exception of a slight delay in the operation of the thruster. This delay allows the canopy to slide back and disengage the canopy hooks before the thruster unit is actuated.

Jettisoning the canopy in this manner also initiates ejection of the seat from the airplane. This is accomplished by the telescopic link connected between the canopy and the canopy-seat ejection interlock mechanism. As the canopy separates from the airplane, the canopy-seat ejection interlock mechanism is actuated and pulls the sear pin from the catapult initiator.

Canopy ejection, where arming of the pilot's ejection seat is undesirable, may be performed by the ground crew by pulling the external CANOPY EMER JETT handle on the right side of the fuselage, below the upper longeron, near the forward end of the canopy (fig. 9). The pilot may also jettison the canopy by means of an internal CANOPY EMER JETT handle on the right instrument wing panel (fig. 2). Either operation

activates an initiator mounted under the canopy bow which fires the
canopy remover and thrust-augmentation-unit initiators. A mechanical
interlock linkage connected to the ballistic-rocket seat-catapult
initiator prevents the initiator from firing (figs. 8(a) and 8(b)).
Although the catapult initiator is not fired at this time, an interlocking
linkage arrangement takes up the slack in the initiator firing linkage so
that actuation of the ejection-seat handles will pull the sear pin from
the initiator.

The external CANOPY EMER JETT handle (fig. 9) is not equipped with
an extension lanyard. Therefore, extreme care must be taken to insure
that no part of the body is directly over the canopy while the handle is
being pulled.

> NOTE: If the ejection handles on the seat are raised, do not pull
> external CANOPY EMER JETT handle until ejection-seat
> initiator hose has been cut; the seat will also be ejected
> when the canopy is jettisoned.

Access to cut the ejection-seat initiator hose is accomplished by
removing the access panel located on the top of the fuselage just aft of
the canopy (figs. 9, 10(a), 10(b), and 10(c)). The hose "cut" point is
painted red.

PARACHUTE ASSEMBLY

Either a 24-foot or a 28-foot pilot's parachute assembly may be
installed in the ejection seat. The pilot's parachute is packed in a
Fiberglas container which is secured in the ejection seat by the lap-
belt restraint-release assemblies on each side of the seat, by the
shoulder-harness straps attached to the headrest, and by the parachute-
release pin on the lower side of the headrest (fig. 5).

The parachute assembly consists of a 24-foot or a 28-foot parachute
canopy, a 30-inch pilot parachute (attached to the main chute apex), a
canopy skirt bag (used to stow the canopy skirt and suspension lines), a
riser assembly (including the upper-pack retaining webbing and adjustable
shoulder-harness straps), an emergency beacon, and two Fiberglas
containers.

PILOT'S EMERGENCY-RADIO BEACON

The pilot's emergency-radio beacon, installed in the pilot's parachute
container, is provided to assist in locating the pilot after he has
ejected from the airplane. The radio-beacon installation is composed of
a radio transmitter, antenna, and battery.

The transmitter is designed to transmit, at 244.3 megacycles, a
pulse-modulated carrier signal which can be detected by ground receiving

stations and search aircraft within a range of approximately 80 miles during descent when the antenna is in a vertical position, and between 15 and 20 miles when the antenna is lying on the ground. The zero-altitude broadcast range can be increased to about 50 miles by attaching a telescopic antenna that is stowed with the radio beacon. Operation of the transmitter is initiated when the lanyard switch plug (attached to the right parachute riser) is pulled from the transmitter and the flexible antenna (also attached to the right parachute riser) is extended upon deployment of the parachute. A 4-hour continuous-operating combination battery supplies three output voltages to power the radio transmitter.

X-15 A/P 22S-2 HIGH ALTITUDE, FULL PRESSURE SUIT ASSEMBLY

DESCRIPTION

The A/P 22S-2 X-15 assembly is a complete extreme-altitude, full pressure flying outfit consisting of high-altitude coveralls, integrated "g" suit, gloves, full pressure helmet, pressure-suit controller, demand oxygen regulator, "g" suit controller, emergency-oxygen supply, pressure socks, sacrifice garment, integrated torso harness, and fabric and exterior boots (figs. 11 and 12). It is normally worn over standard long underwear, but thermal or insulated underwear may be used.

The assembly operates routinely from the aircraft's oxygen and vent-air systems. During normal operation the suit does not become pressurized, but in the event cabin altitude exceeds 36,500 feet, an automatic aneroid operation pressurizes the assembly to approximately 35,000 feet. An emergency back-pan assembly containing an oxygen package is utilized operationally.

FUNCTION

The purpose of the A/P 22S-2 X-15 assembly is to provide an environment for the pilot which will meet all physiological requirements under various normal flight and emergency conditions at extreme altitudes. The suit accomplishes this function in the following manner:

    Provides an adequate oxygen supply with correct breathing pressure.
    Maintains a coverall pressure of about 175 millimeters of mercury
       (±7) (34,600 to 36,200 ft).
    Provides "g" protection from 3g to 12g.
    Provides good exposure protection in water or on land.
    Allows freedom of movement for required activities.
    Enables the pilot to remain with the aircraft and in full control
       of it, or to eject, if necessary.
    Maintains thermal balance during elevated temperatures through the
       use of an integrated ventilation system and a heat-reflective
       coverall.

B-52 MODIFICATION

GENERAL DESCRIPTION

Two B-52 airplanes (S/N 003 and S/N 008) have been modified as carrier airplanes for the X-15 research program (fig. 13). The modification consists of the following installation and/or rework of system units and components to support the X-15 airplane on the ground and during a flight mission:

An AN/APN-81 Doppler radar system is installed to provide ground speed and drift-angle information to the stable platform in the X-15 airplane.

A closed-circuit television system is installed to monitor the pre-release and flight operations of the X-15 airplane.

An auxiliary UHF communication system is installed to provide additional communication channels.

The AN/AIC-10 interphone system is changed to provide an AUX UHF position.

Liquid-oxygen top-off tanks are installed in the bombbay area for transferring, to the X-15 airplane tank, the liquid oxygen that is normally boiled off during flight. The carrier-airplane forward-body fuel tanks and mid-body tanks are removed to accommodate this installation. The upper fuel-cell area is reworked to accommodate the installation of 18 storage cylinder tanks. These tanks are used to supply system pressure for the liquid-oxygen top-off tanks, pressure-suit ventilation, and the launch emergency-release system.

The right main wing tank is removed to allow for carrier pylon-tie fittings and supports in the front wing spar and rear wing spar. The fittings are used to fasten the carriage pylon to the carrier airplane. The right and left inboard flaps are locked so that the flaps remain in a fixed retract position at all times. The inboard-flap mechanism is disconnected, and the flaps are bolted to the flap tracks. The carriage pylon is bolted to tie fittings between the right inboard engine nacelle and the carrier fuselage. System components from the carrier fuselage terminate at quick-disconnect fittings in the pylon. The purpose of the pylon is to carry the X-15 airplane from the ground and during flight, and to supply the necessary system functions from the carrier airplane to the X-15.

The launch operator's station replaces the electronics countermeasure compartment. The launch operator monitors the liquid-oxygen top-off system control panel, radio and television equipment, and radar installation for the research program. Provisions are included for telemetry equipment.

The escape hatch is changed to allow the operator to hold the hatch open during ground servicing operations. A viewing window is added to the right side for observation, and the portable oxygen bottle is relocated below the window.

Helium gas is used to pressurize the contents of the top-off tanks for transfer or jettison purposes. Nitrogen gas is used for pneumatic actuation of the liquid-oxygen control valves.

Breathing oxygen is available to the technical crew members at all times. Also, oxygen is tapped from the carrier oxygen system to supply the X-15 pilot with breathing oxygen until flight release.

Pressure-suit ventilation is available to the X-15 pilot. The system is supplied by nitrogen-storage cylinders in the fuel-cell area and is controlled by the research pilot according to his needs. Electrical connections are provided so that carrier-airplane power can be used for testing the X-15 airplane electrical equipment and instruments. Connections also allow the launch operator's panel to accomplish its intended function.

The X-15 airplane is released by mechanical linkage, in the pylon, operated normally by hydraulic pressure and operated in case of an emergency by emergency pneumatic pressure augmented by an accumulator.

Propellant jettison lines, leading aft and overboard in the carrier, jettison liquid oxygen into the atmosphere in case of an emergency.

GROUND-RESCUE PROCEDURE

X-15 PILOT REMOVAL

The following steps are to be taken when the <u>ejection handles are not raised</u>, and under <u>NORMAL recovery conditions</u>:

1. Upon arrival at the aircraft, ascertain condition of X-15 pilot.

2. Verify that ejection-seat handles are in the down and locked position. (Position of ejection-seat handles can be checked by looking through left-side canopy windshield at the mirror mounted on the circuit-breaker panel below the right-canopy longeron (figs. 14(a) and 14(b)).

3. Open canopy upon pilot's direction (fig. 9).

4. Pilot completes postflight checklist and operations to secure X-15 aircraft.

5. Install ejection-seat and emergency restraint-release-handle safety pins (fig. 5).

6. Release parachute risers and parachute "D" ring retainer-strap quick disconnects (fig. 7(a)).

7. Remove lap-belt buckle (fig. 7(b)).

8. Release foot-manacle restraints if pilot's feet are locked in stirrups (fig. 7(c)).

9. Disconnect pilot's emergency-oxygen actuation cable. (Pull "Green Apple" to arm emergency-oxygen system if high concentration of ammonia is present.)

10. Open visor of pilot's helmet. (Visor is to remain closed if ammonia is present.)

11. Turn X-15 oxygen-selector valve to OFF and pressure-suit vent valve to CLOSED (fig. 5).

12. Release pilot's personal-lead quick disconnect (fig. 7(d)).

13. Assist pilot in exit from cockpit.

The following steps are to be taken when the <u>ejection handles are raised</u>, and for an **EMERGENCY** or damaged X-15 recovery:

    1. Upon arrival at the aircraft, ascertain condition of the X-15 and the pilot.

    2. Remove the access panel on the top of the fuselage, just aft of the canopy. Cut the ejection-seat initiator hose at the centerline just forward of the bulkhead. The hose "cut" point is painted red (figs. 10(a), 10(b), and 10(c)).

    3. Push the yellow button to unlatch the cover door on the right side of the forward fuselage. Pull out the manual release handle and rotate the handle up and back to unlatch the canopy (fig. 9).

    4. Lift the canopy at the forward end, then push up and aft until the canopy completely separates from the airplane.

> NOTE: If time does not permit, or if the canopy cannot be opened with manual release, jettison the canopy by pulling the yellow "T" handle just forward of the manual release handle (fig. 9).

WARNING   There is no extension cable on the jettison handle. Keep all personnel clear of the canopy-ejection path.

    5. Pull pilot's "Green Apple" to arm the emergency-oxygen system.

    6. Simultaneously push both of the ejection handles outboard and down to the full stowed position.

> NOTE: If both of the ejection handles will not move full down, pull the restraint emergency release handle to free the pilot of all restraints (figs. 5 and 6). Then, disconnect the parachute "D" ring retainer strap, both parachute risers, and the lap-belt buckle and remove the pilot from the seat.

WARNING   Do not stand above the seat when the restraint emergency-release handle is pulled; the head-rest is ejected straight up with explosive force.

    7. Release parachute risers and parachute "D" ring retainer-strap quick disconnects (fig. 7(a)).

    8. Remove lap-belt buckle (fig. 7(b)).

9. Release foot-manacle restraints if pilot's feet are locked in the stirrups (fig. 7(c)).

10. Release pilot's personal-lead quick disconnect (fig. 7(d)).

11. Remove pilot.

SECURING X-15 AIRCRAFT

Under normal conditions, securing the X-15 and shutting down the systems will be accomplished by the pilot as specified in the pilot's check list.

In an emergency, ground-rescue personnel will secure the X-15 as follows:

1. Place both APU switches in the OFF position (fig. 15).

NOTE: If the APU's fail to shut down when the switches are off, turn the emergency battery switch on, while the APU switches remain in the off position. After the APU's shut down, place the emergency battery in the off position.

2. Place the pressurization-system selector control handle in the VENT position if the propellant tanks are pressurized and the selector control handle is in the pressurized position (figs. 14(a) and 14(b)).

NOTE: Placing the selector control handle in the VENT position will vent the pressurizing gases from the propellant tanks and will be accompanied by a loud noise made by the gases escaping from the propellant-tank vent valves.

3. If the pressurization-selector control handle is in the JETTISON position, do not move it.

WARNING Do not place control handle in the JETTISON position. This will dump any remaining fuel at the aft of the aircraft causing a highly explosive mixture as the ammonia, oxygen, and hydrogen peroxide combine.

4. Evacuate the area, pending arrival of fire-fighting and rocket-systems personnel.

X-15/B-52 MATED CONDITION

REMOVING AND/OR RESCUING THE X-15 PILOT

Removal and/or rescue of the X-15 pilot can be accomplished by several methods. The procedure to be used in the various methods is:

<u>NASA travel-all 4012</u>
 (fig. 16(a))
Follow steps for normal pilot removal (page 19) or emergency pilot removal (page 20), depending on rescue conditions.

<u>Ladders</u> (fig. 16(b))
 NASA 3/4-ton rescue truck
 Fire department R-2
  rescue rig
Follow steps for normal pilot removal (page 19) or emergency pilot removal (page 20), depending on rescue conditions.

<u>Fire trucks</u>
 Class O-11 A and B
  (fig. 16(c))
 Class O-6 (fig. 16(d))
 R-2 rescue rig
  (fig. 16(e))
Follow steps for normal pilot removal (page 19) or emergency pilot removal (page 20), depending on rescue conditions.

REMOVING THE B-52 CREW

Removal of the B-52 crew will be accomplished by fire-department personnel or the B-52 ground crew with the following procedure:

Open the entry and rescue hatch at the nose of the aircraft by pushing handle lock-release button and rotating handle to unlock hatch and lower to full-down position.

Stand by to assist crew if needed.

X-15/B-52 EMERGENCY CONDITIONS AND PROCEDURES

PRIOR TO TAKEOFF

In emergencies occurring before takeoff, the procedure is as follows:

FIRE OR EXPLOSION

B-52 crew	Stop, shut down engines, abandon airplane.
X-15 pilot	Open canopy, abandon aircraft with aid of rescue personnel.
Firemen	Water coverage, as required. Aid in rescue of X-15 pilot and aid B-52 crew.
Lake controller	Abandon area. Keep area clear through radio and P.A. announcements.

FIRE WARNING

B-52 crew	Stop, shut down engine, prepare to abandon airplane.
X-15 pilot	Open canopy, prepare to abandon with aid of rescue personnel.
Firemen	Prepare to water-soak engine compartment. Aid rescue personnel and X-15 pilot.
Lake controller	Initiate action of firemen and rescue personnel by radio and P.A. announcements. Ascertain when emergency is secured. Direct subsequent action.

H$_2$O$_2$ OR APU COMPARTMENT HOT WARNING

    B-52 crew — Stop B-52 heading into wind, if possible. Shut down engines, abandon airplane.

    X-15 pilot — Radio call intention to jettison H$_2$O$_2$. Proceed to jettison, contingent upon wind direction. Open canopy in preparation for abandoning airplane.

    Firemen — Water-flush H$_2$O$_2$. Aid rescue personnel and X-15 pilot.

    Lake controller — Initiate action of firemen and rescue personnel via radio and P.A. announcements. Keep area clear.

PROPELLANT SPILLAGE

    Same procedure as for FIRE WARNING, except firemen flush propellant.

IN FLIGHT

For in-flight emergencies, the B-52 launch commander verifies conditions and initiates proper procedures in accordance with the following:

    <u>X-15 hazardous to B-52</u> — Launch X-15

(This action may be preceded by X-15 pilot ejection if X-15 is hazardous to X-15 pilot.)

    <u>B-52 hazardous to X-15</u>

        Below critical altitude — Eject X-15 pilot / Eject B-52 crew

        Above critical altitude — Launch X-15 / Eject B-52 crew

    <u>X-15 emergency not immediately hazardous</u> — Return to base for landing. Jettison propellants.

    <u>B-52 emergency not immediately hazardous</u> — Return to base for landing. Jettison propellants.

See flight check list for emergency launch and ejection procedures.

FIRE WARNING

X-15 pilot	Call "fire-warning light on".
Launch operator } Chase pilot	Report condition of aft fuselage. Continue to monitor.
X-15/B-52 pilots	Prepare for emergency launch.

H_2O_2 OR APU COMPARTMENT HOT WARNING

X-15 pilot	Report condition.
Launch operator } Chase operator } B-52 co-pilot	Report condition of X-15. Look for evidence of smoke.
Launch operator	APU compartment heat off.
X-15 pilot	Jettison H_2O_2 from hot system. Prepare for emergency launch.

LANDING

For landing emergencies, the procedure is as follows:

B-52 pilot	Radio-call nature of emergency and location of probable emergency landing.
Lake controller	P.A. call "All emergency equipment proceed to landing area. Follow me." <u>Remain clear of possible airplane path</u>.
B-52 pilot	Give periodic status reports. Initiate liquid-oxygen jettison.
X-15 pilot	Initiate propellant jettison.
B-52 pilot	Land airplane, shut down engines, and secure airplane. Crew leave airplane.
Lake controller	Proceed to X-15/B-52 with fire truck and crash truck. Direct maintenance personnel to install pylon safety pin.

Firemen or rescue personnel	Open X-15 canopy and safety seat: 1. Push yellow button to unlatch door (right-hand side). 2. Pull out manual release handle. Rotate handle up and back to unlatch canopy. 3. Lift forward edge of canopy until opened. 4. Activate pilot's emergency oxygen system. ("Green Apple") 5. Install three safety pins or cut initiator hoses as marked on X-15. 6. Assist pilot in disconnecting personal leads and restraining mechanism. 7. Assist pilot from cockpit.
Lake controller	Secure both airplanes and initiate action to recover aircraft from landing site.

X-15 EMERGENCY

The following action by the X-15 pilot may be expected in the specified X-15 emergency:

No nose-gear extension Airplane out of control	EJECT
Loss of visibility	JETTISON CANOPY
Main gear up, nose gear down Failure to jettison lower vertical Asymmetrical flaps	LAND

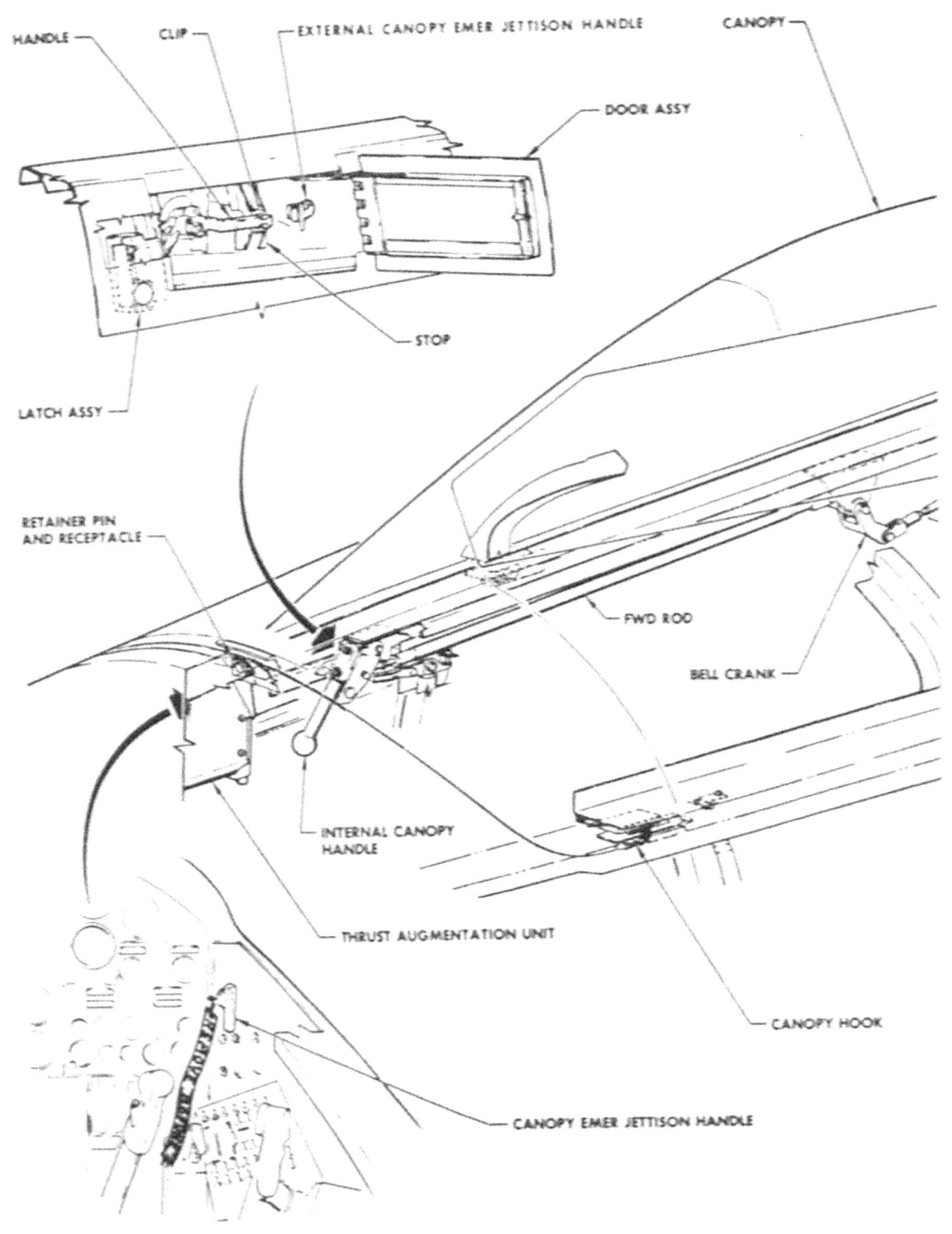

Figure 2.- Cutaway drawing showing internal and external canopy release mechanisms.

THIS PAGE OMITTED.

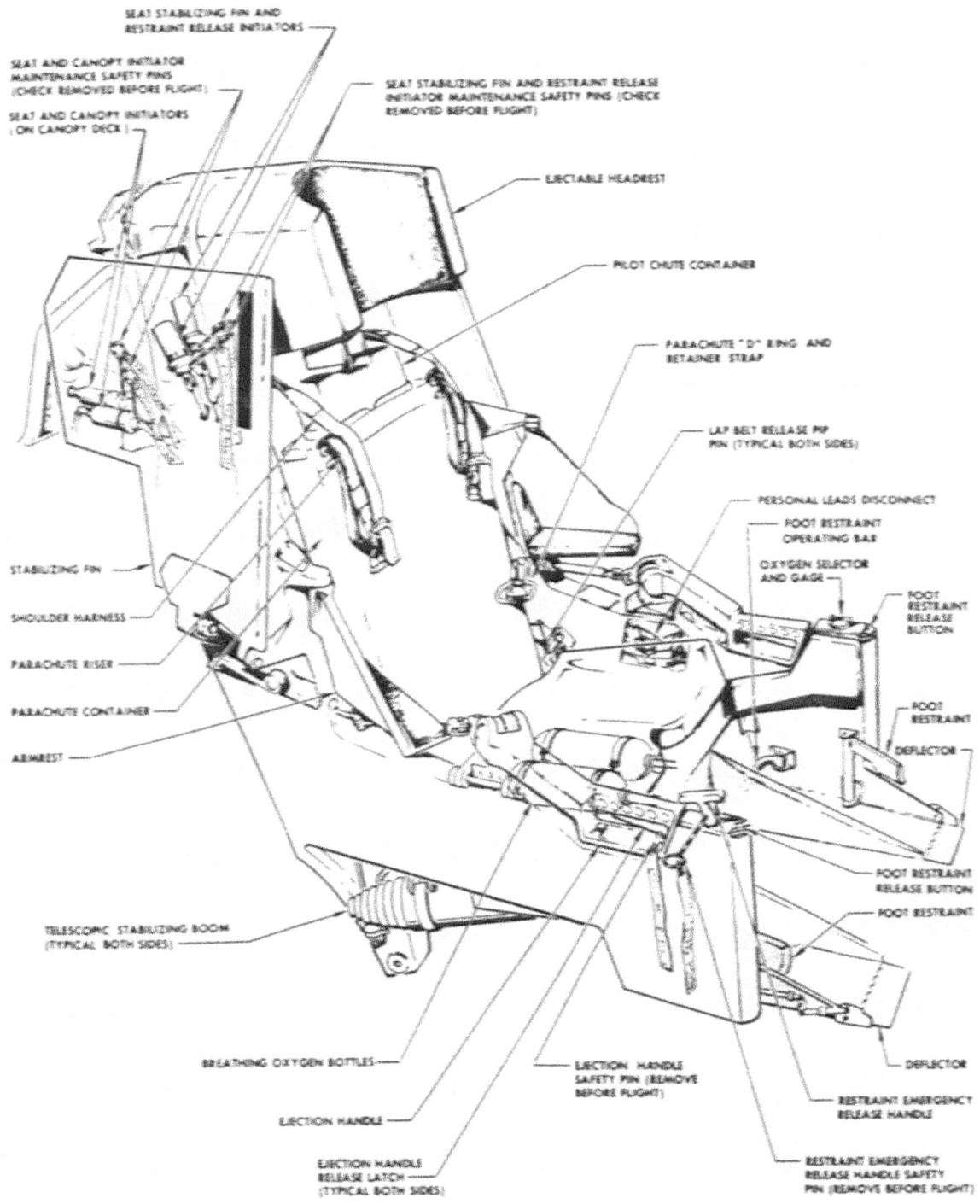

Figure 5.- X-15 ejection seat.

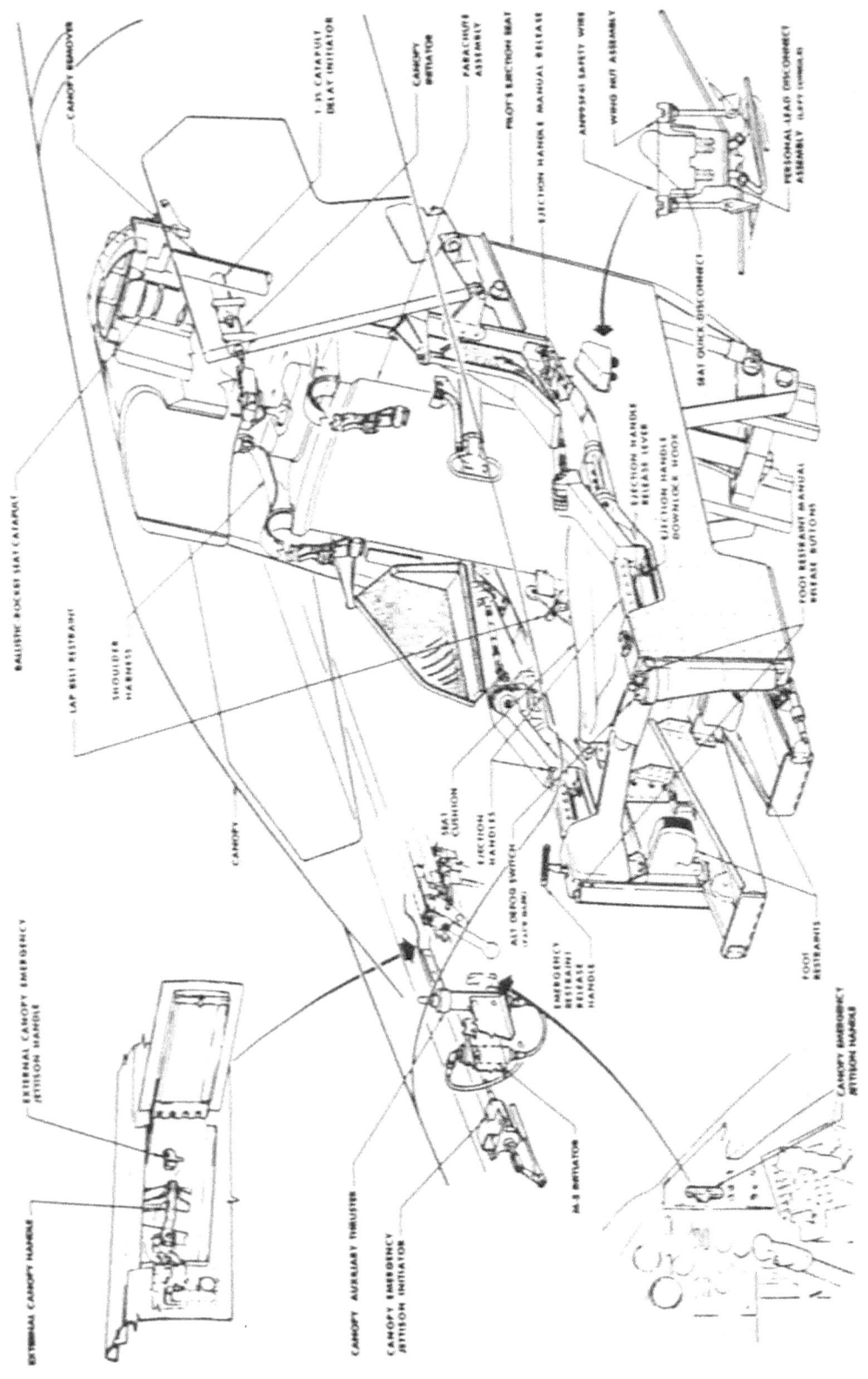

Figure 6.- X-15 seat and canopy.

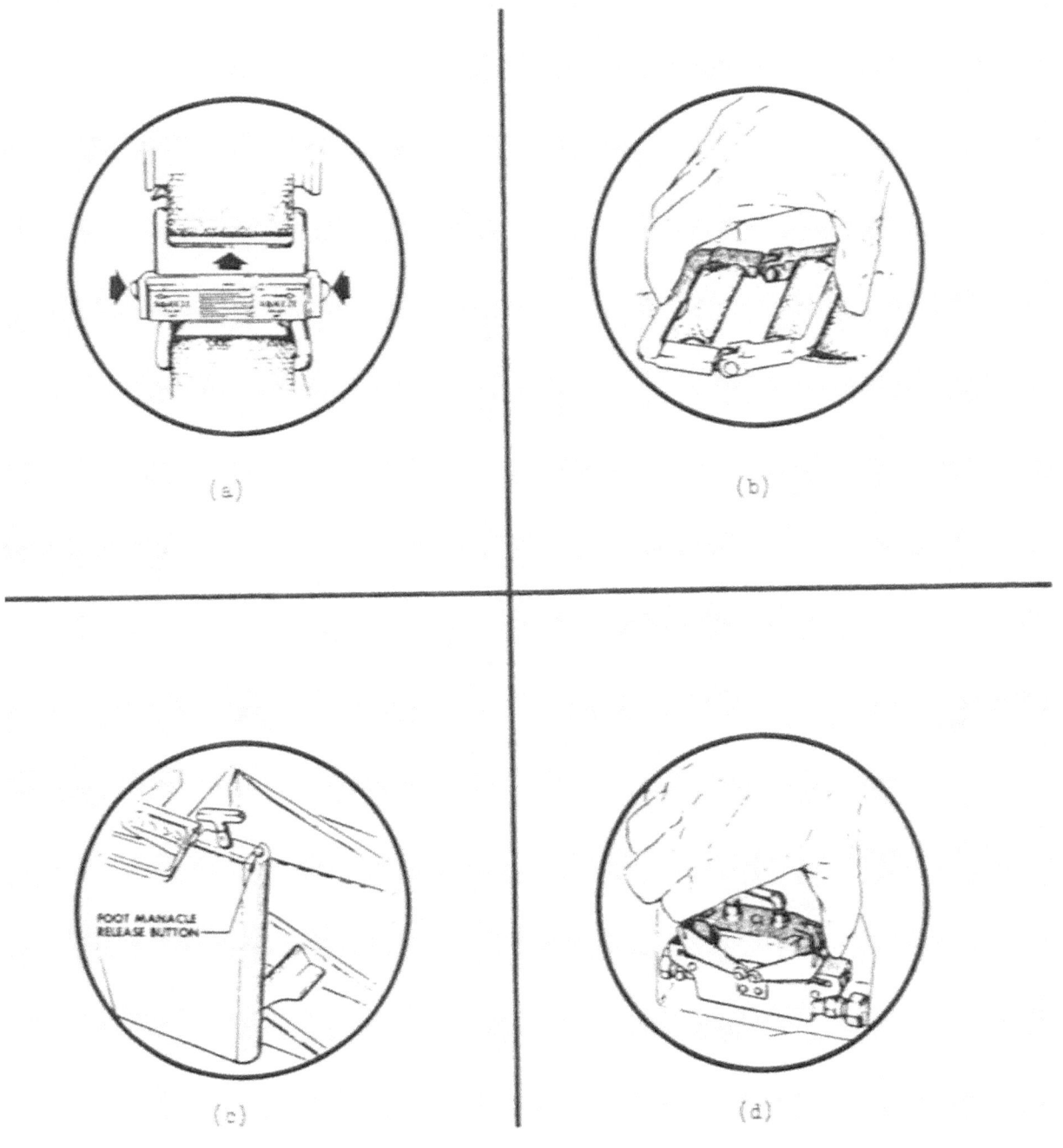

Figure 7.- Pilot-release mechanisms.

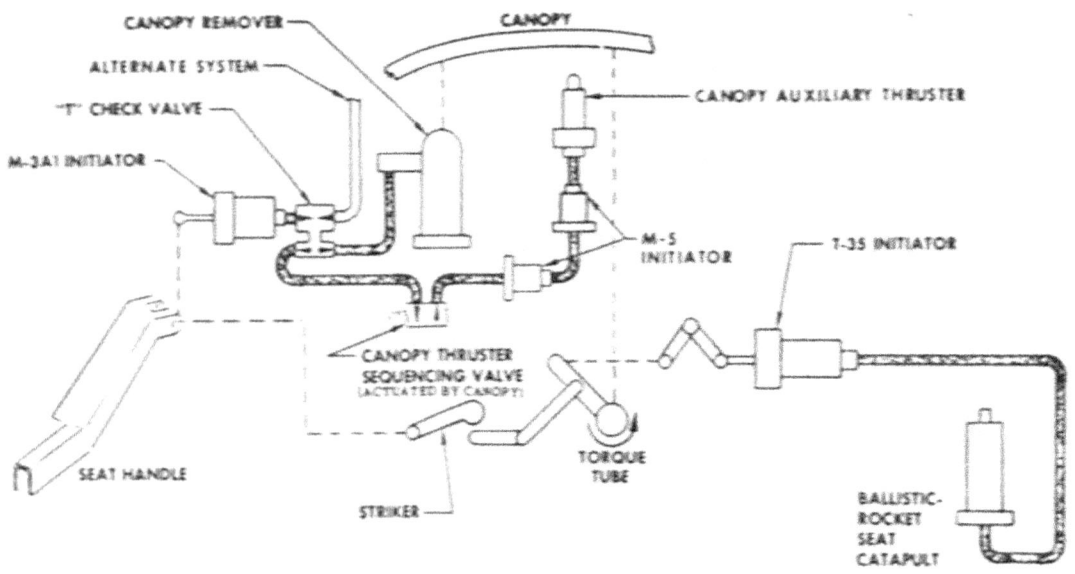

(a) Normal sequence.

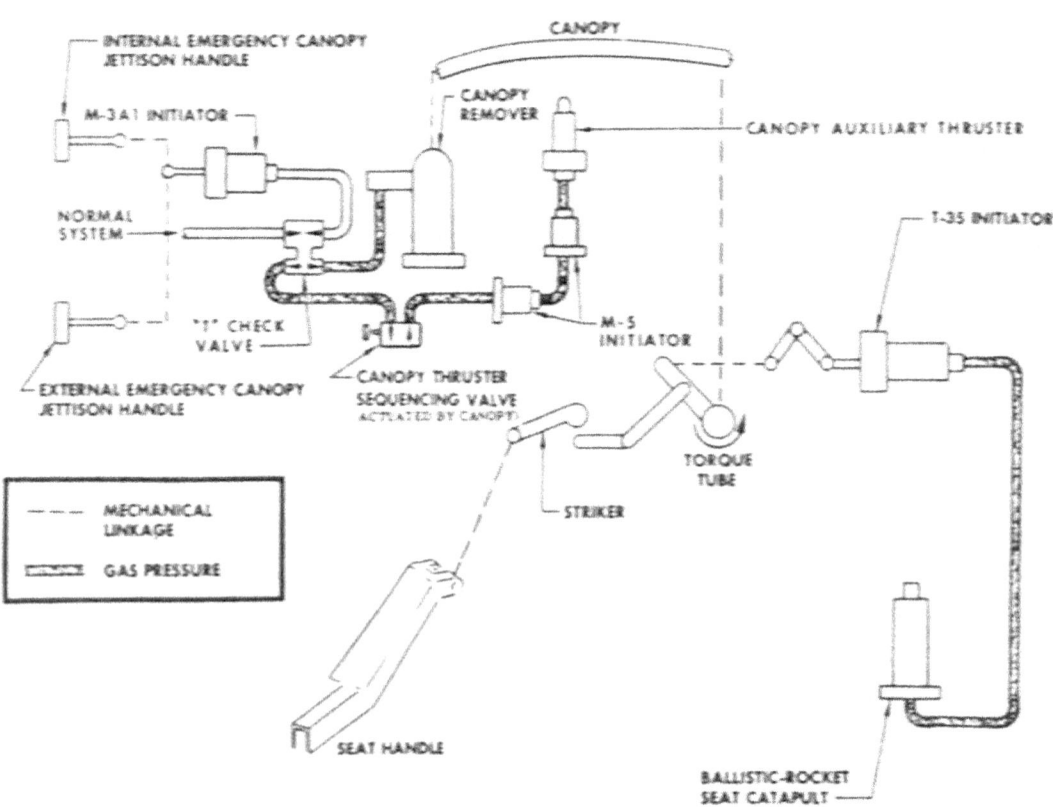

(b) Alternate sequence.

Figure 8.- Canopy-ejection system.

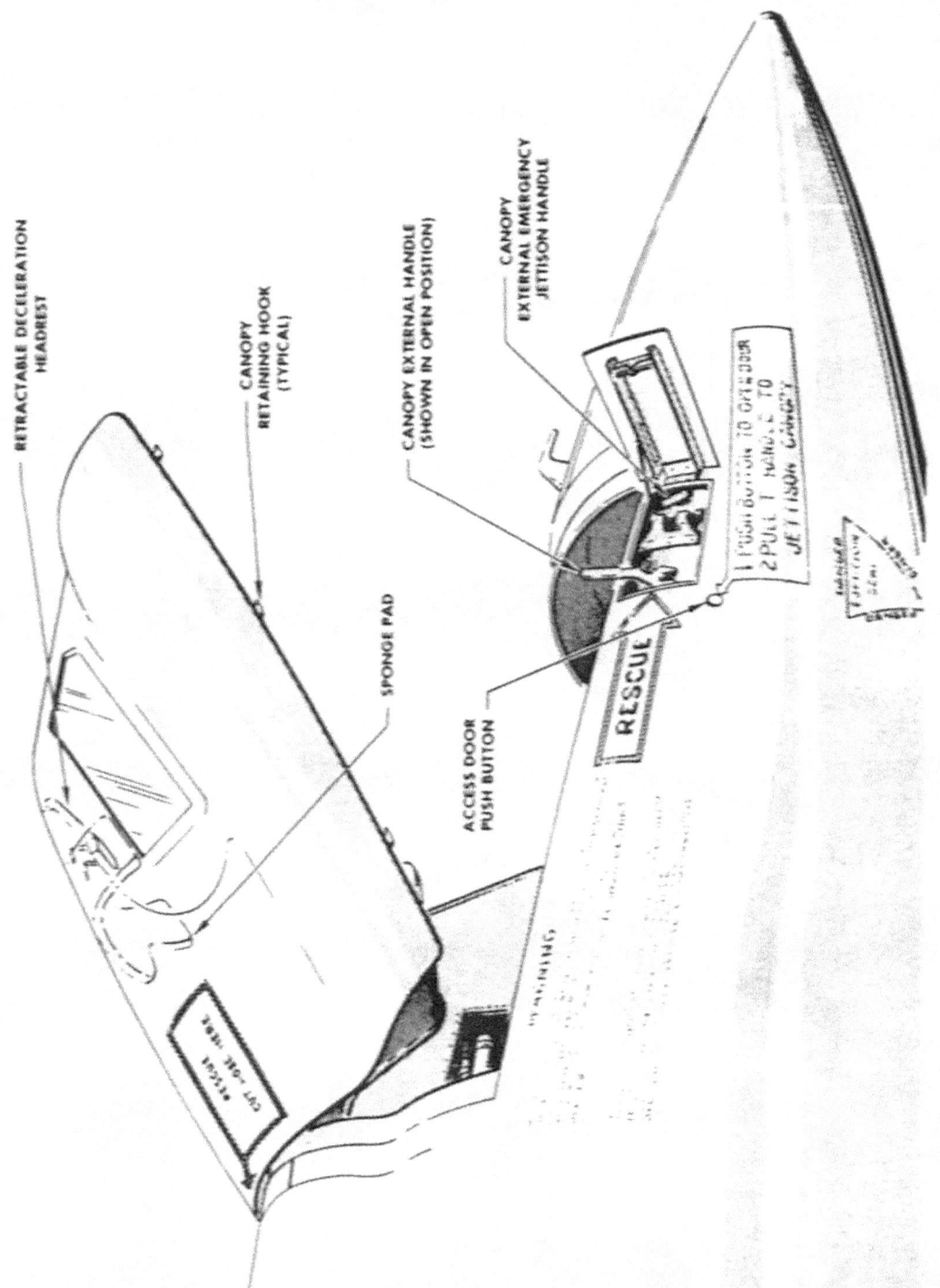

Figure 9.- View of right side of X-15 cockpit area.

(a) Canopy removed.

Figure 10.- Photographs of X-15 ejection-seat initiator hose.

(b) Canopy in place, cover plate removed.

Figure 10.- Continued.

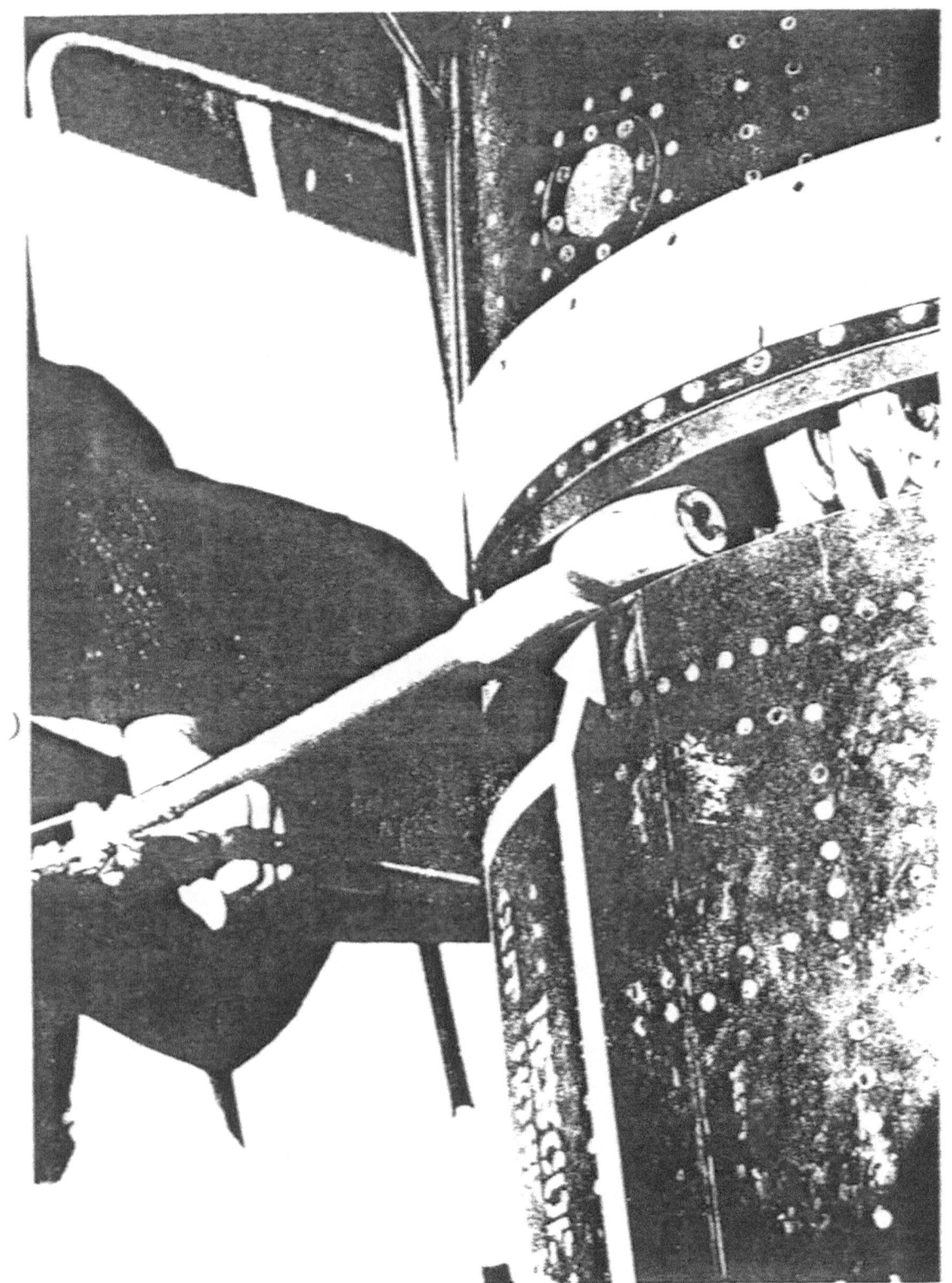

(c) Hose being cut with cutter.

Figure 10.- Concluded.

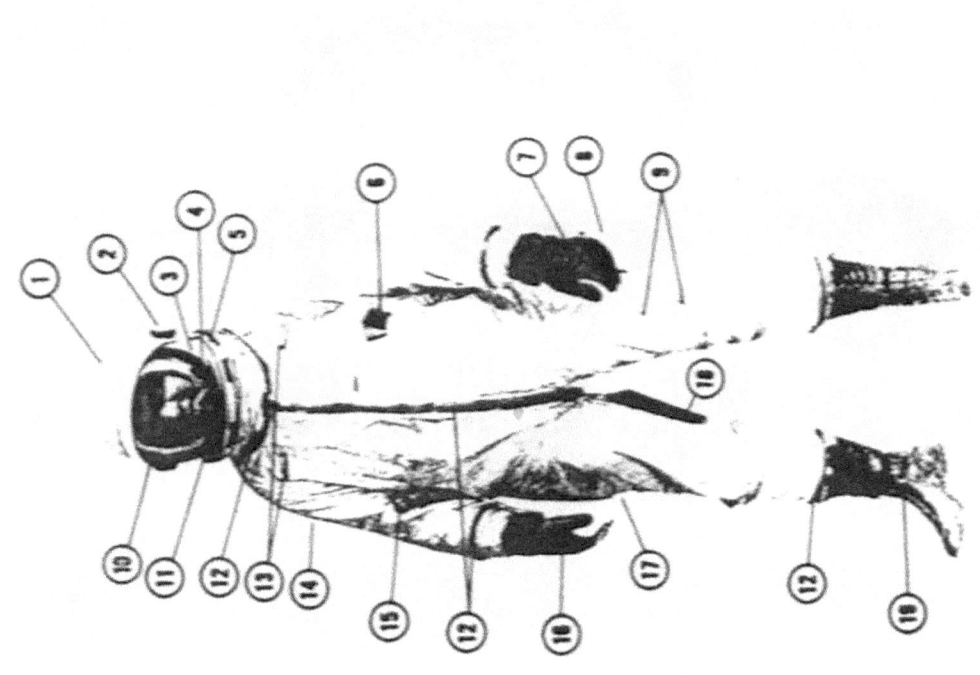

1. Helmet - Fiberglas, rigid construction
2. Visor latch release and oxygen on-off valve
3. Visor - Plexiglas, electrical conductive coating
4. AIC/10 microphone
5. Neck ring, detachable
6. Parachute "D" ring (ripcord) attaching ring
7. Palm restraint strap
8. Personal leads, disconnect (man-to-seat half)
9. Snaps for pilot's checklist
10. Two-way face seal
11. Inflatable visor seal
12. Closure zippers
13. Parachute-riser and shoulder-harness attaching rings
14. Aluminized sacrifice garment
15. "Green Apple"
16. Gloves
17. Lap belt and integrated torso harness
18. Survival knife
19. Exterior boot

Not shown: Back pan - contains 15 minutes of emergency oxygen, g suit valve, pressure reducer, electrical wiring and tubing to route necessary functions from airplane to pilot

Inner X-15 A/P 22S high-altitude, full pressure flying outfit

Figure 11.- A/P 22S-2 X-15 full pressure suit.

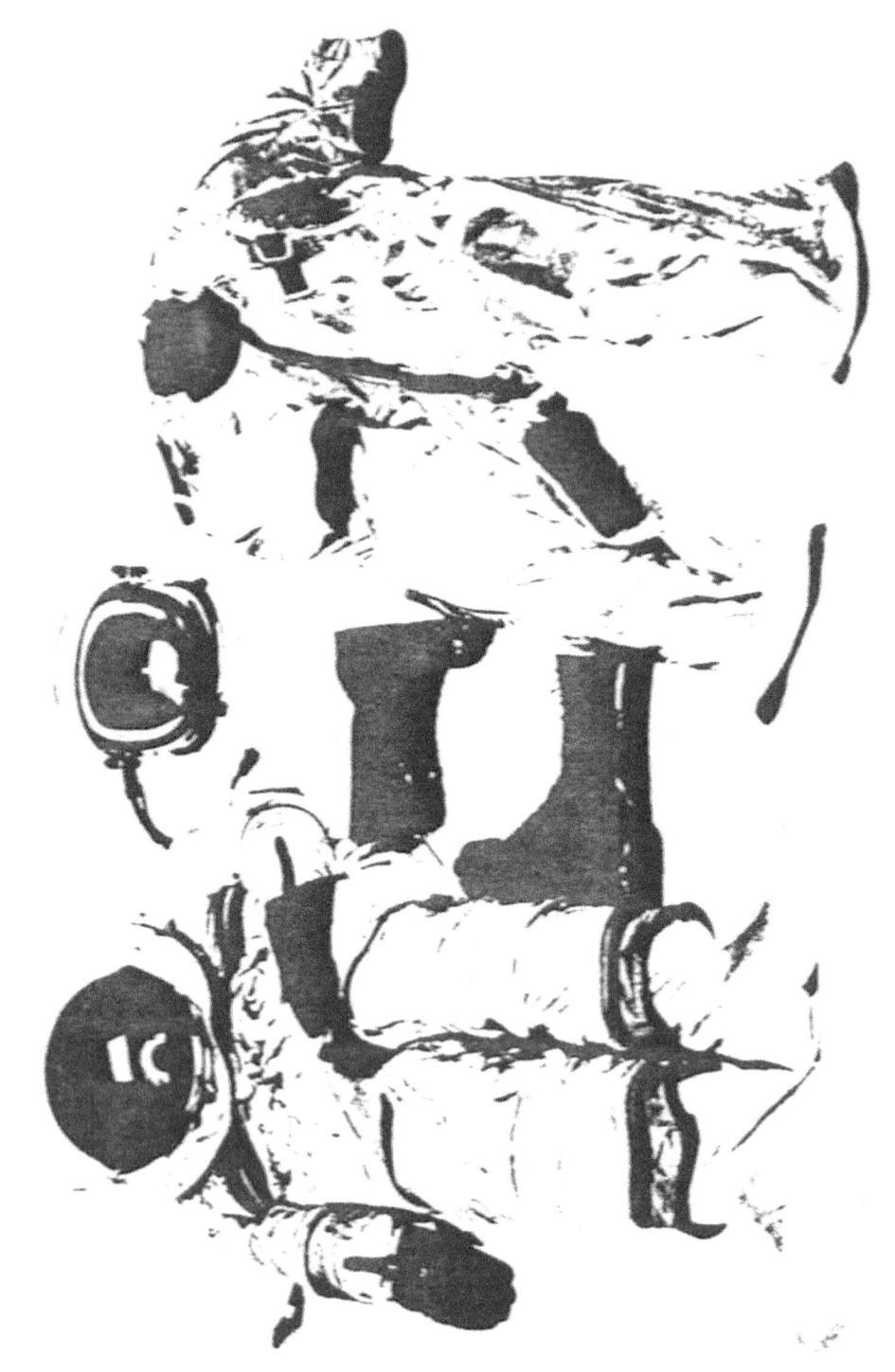

Figure 12.- Layout of A/P 22S-2 X-15 full pressure suit.

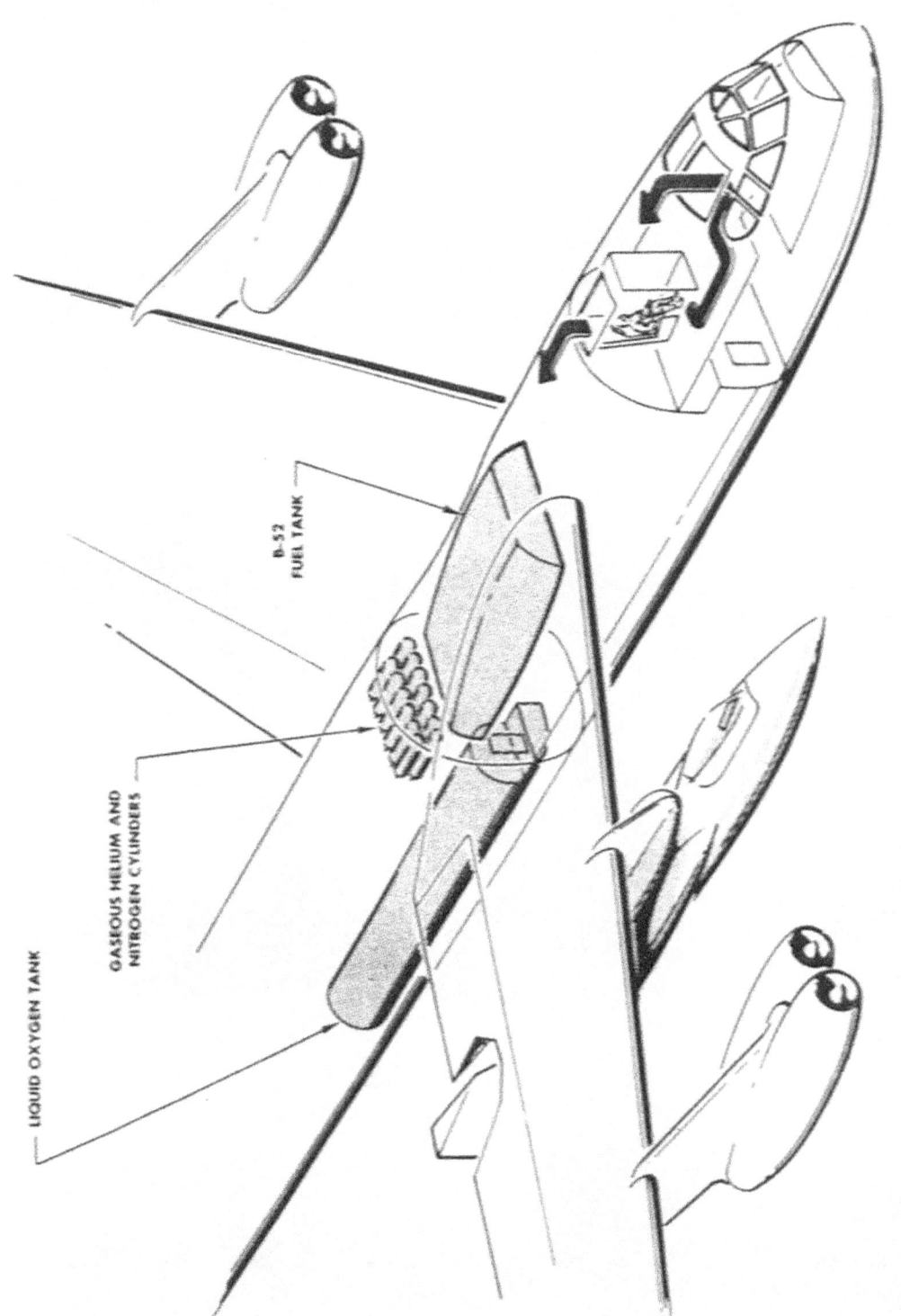

Figure 13.- Modification of B-52 for X-15 program.

(a) Right-hand instrument panel.

Figure 14.- Views of X-15 cockpit.

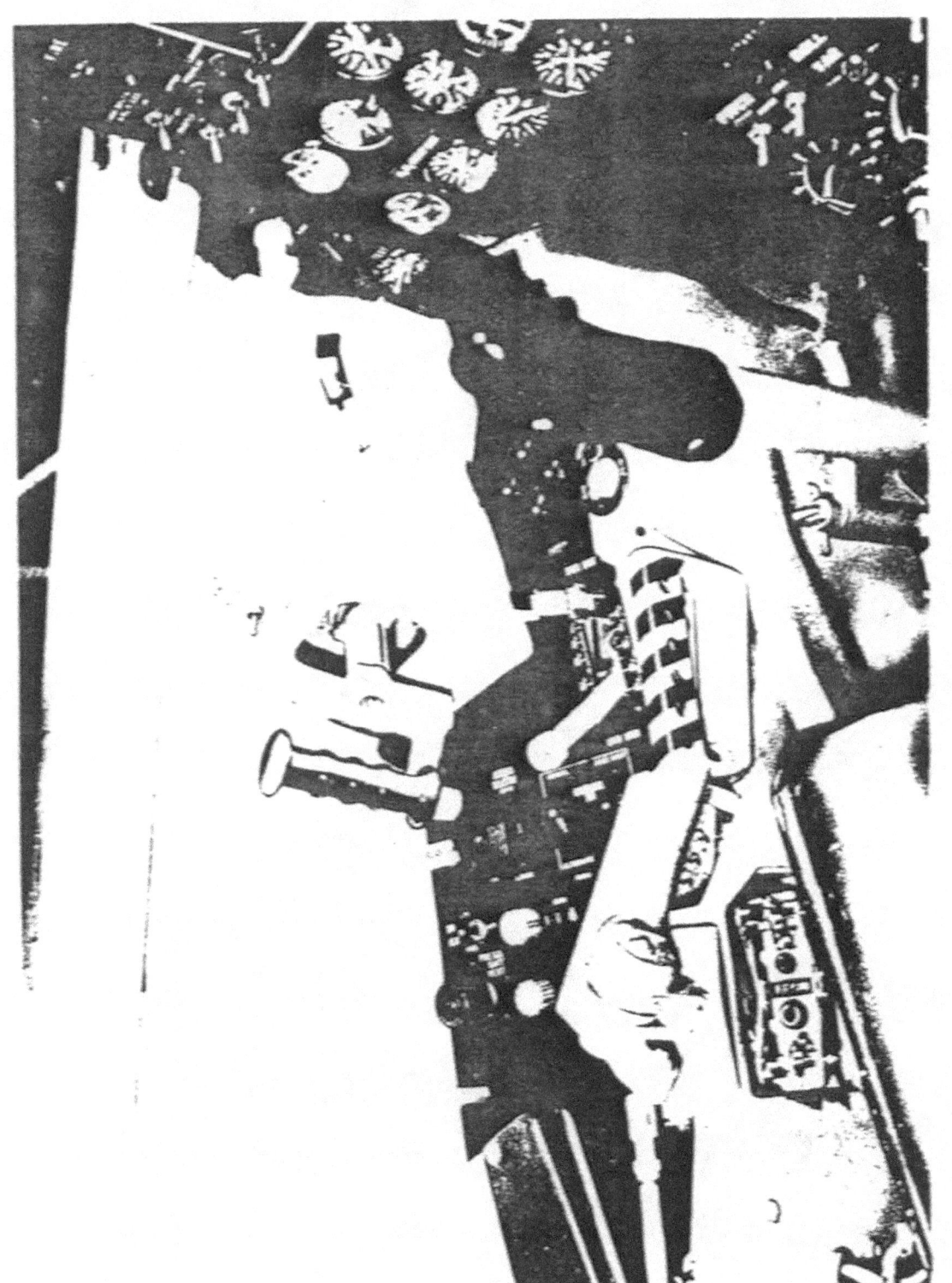

(b) Left-hand instrument panel.

Figure 14.- Concluded.

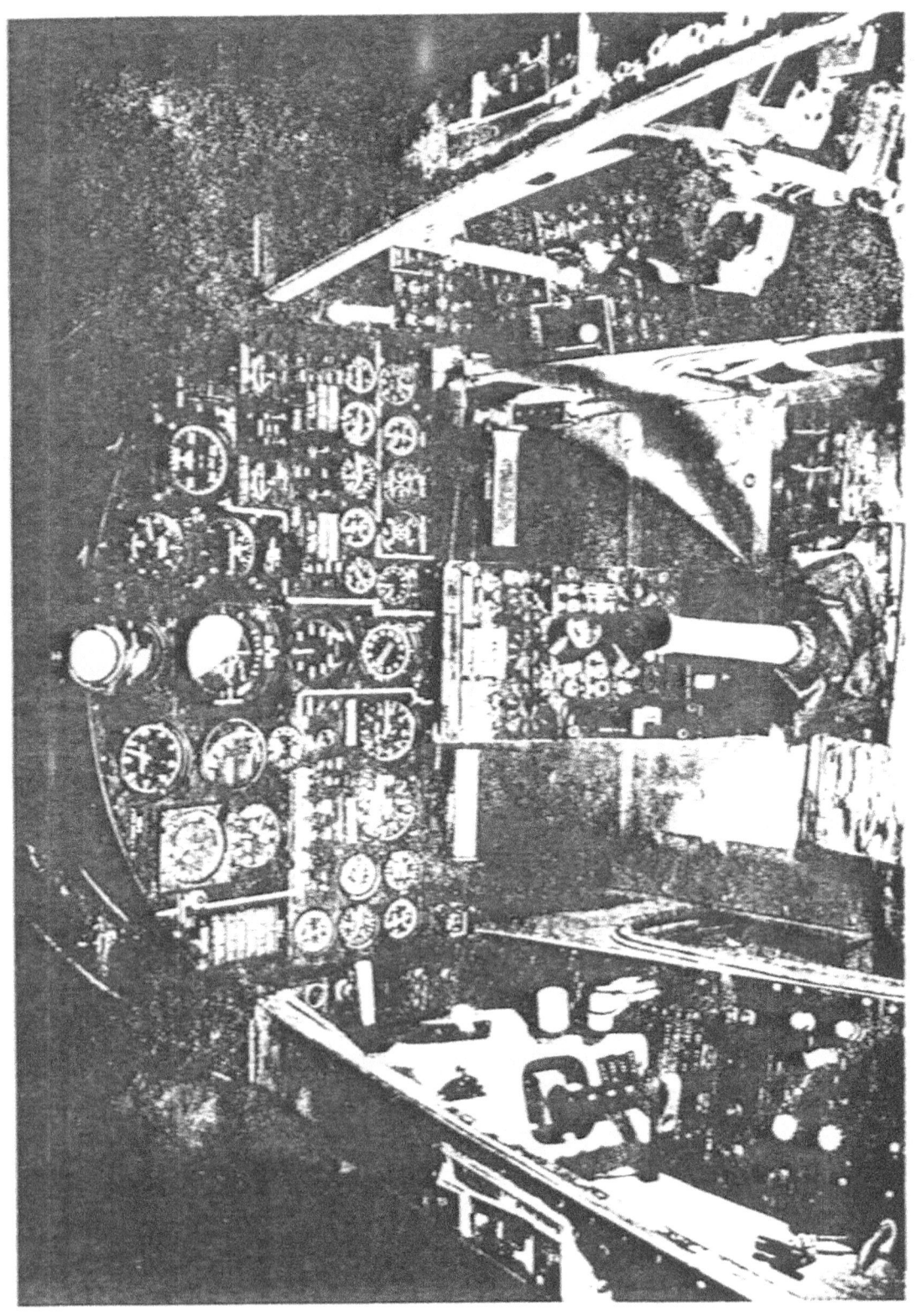

Figure 19.- Interior of X-15 cockpit.

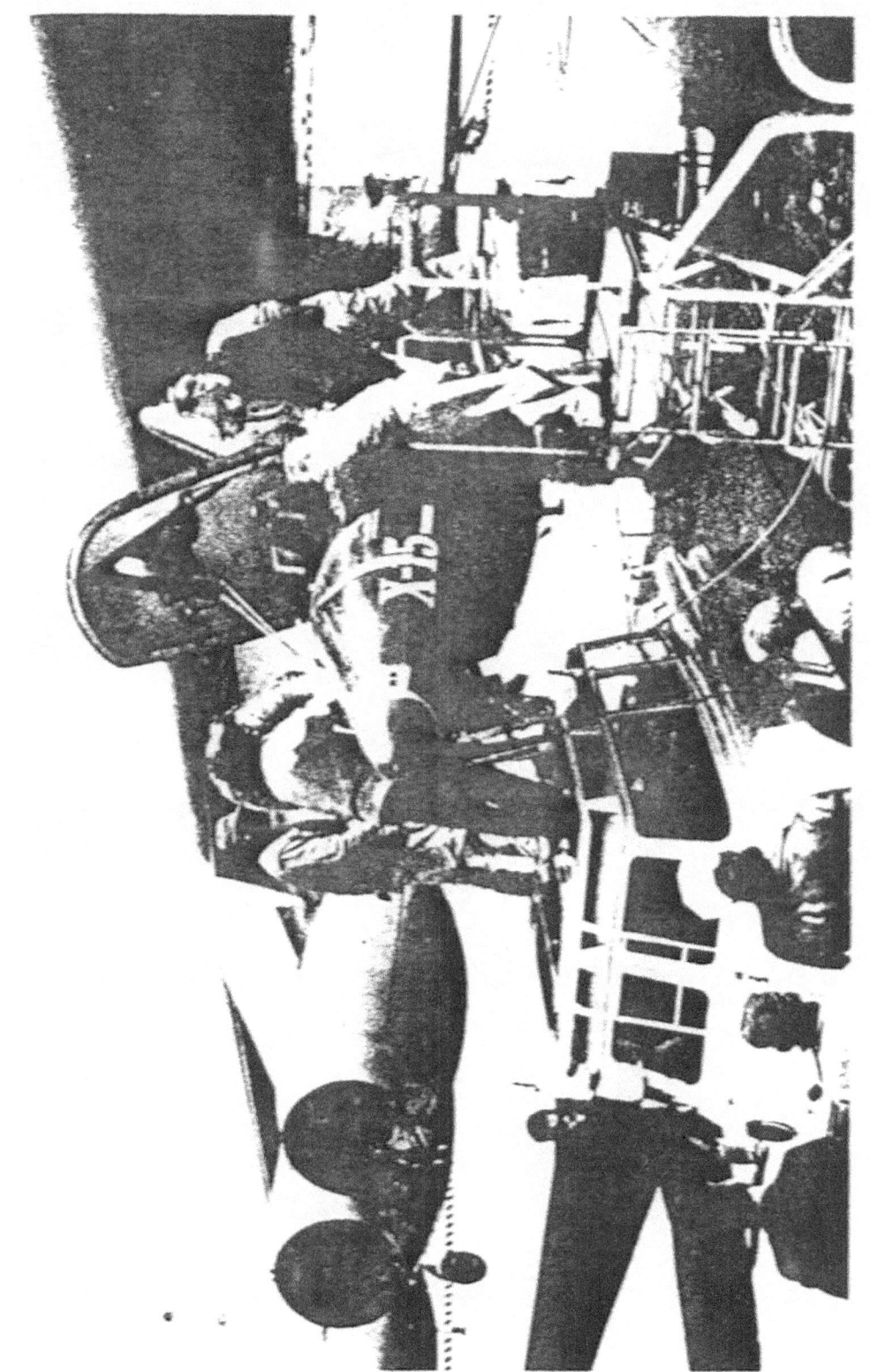

(a) NASA travel-all.

Figure 16.- Rescue of X-15 pilot in X-15/B-52 mated condition.

(b) Ladder.

Figure 16.- Continued.

(c) O-11 type fire truck.

Figure 16.- Continued.

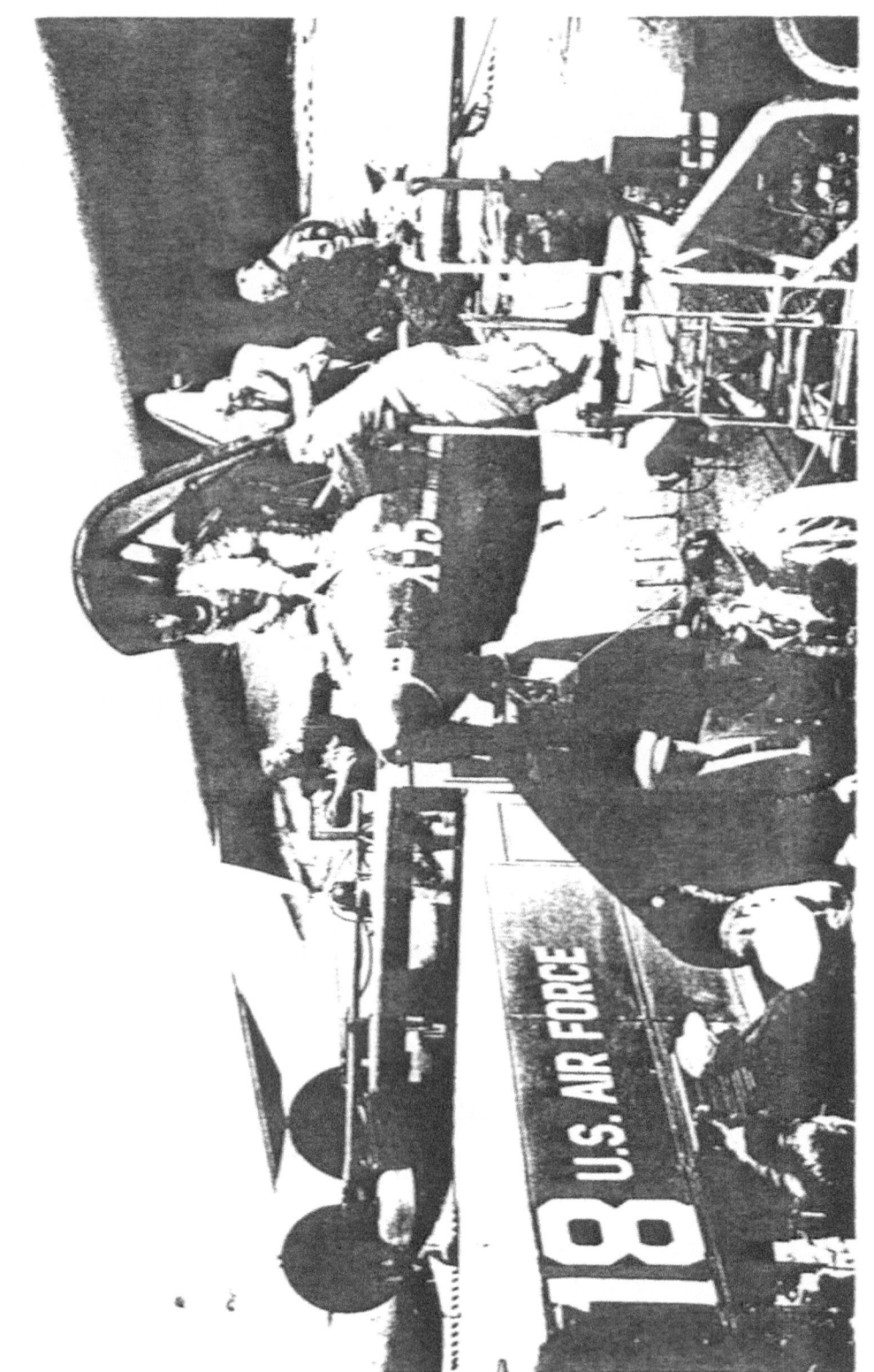

(d) O-6 type fire truck.

Figure 16.- Continued.

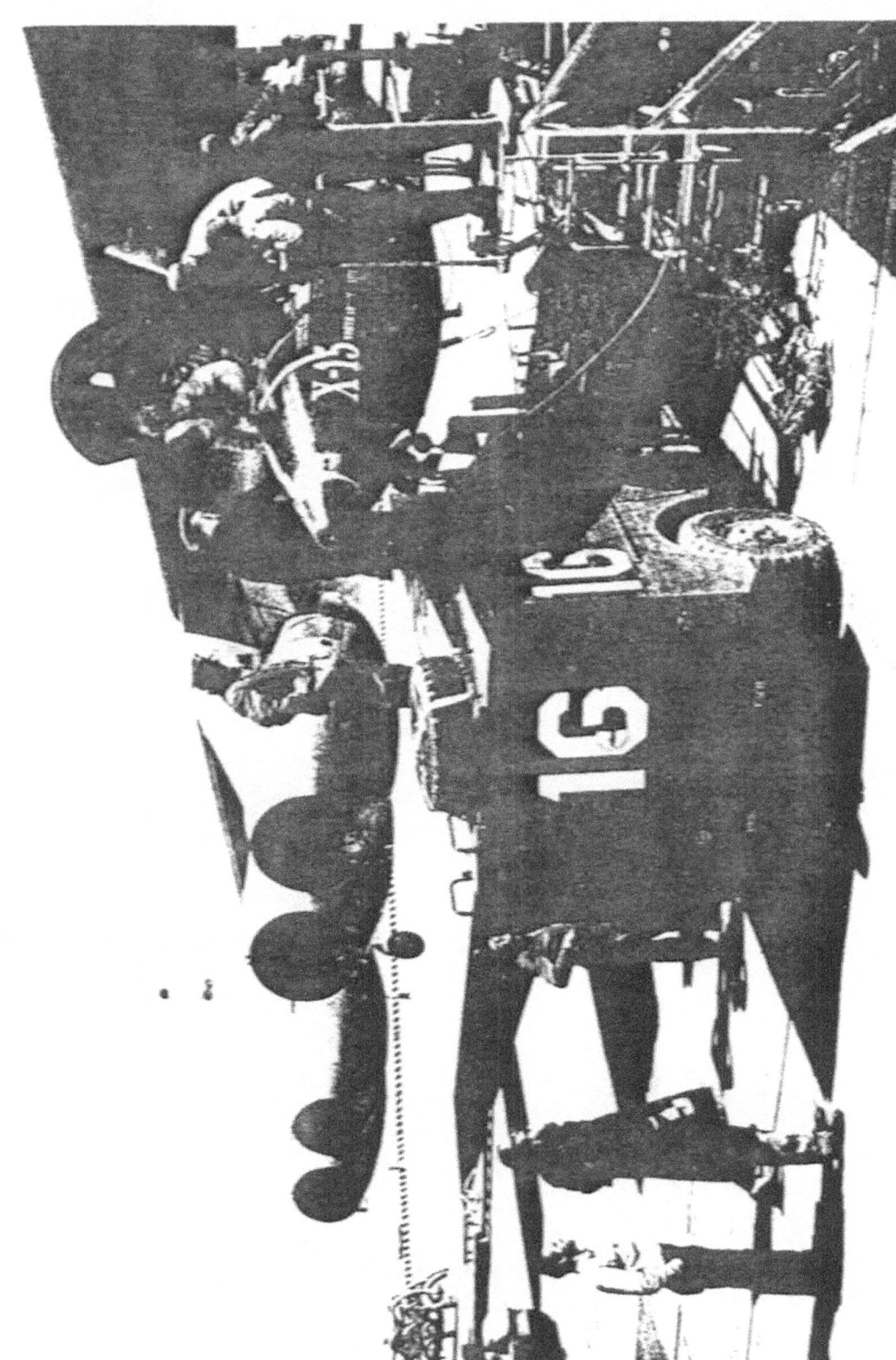

(e) R-2 rescue rig.

Figure 16.- Concluded.

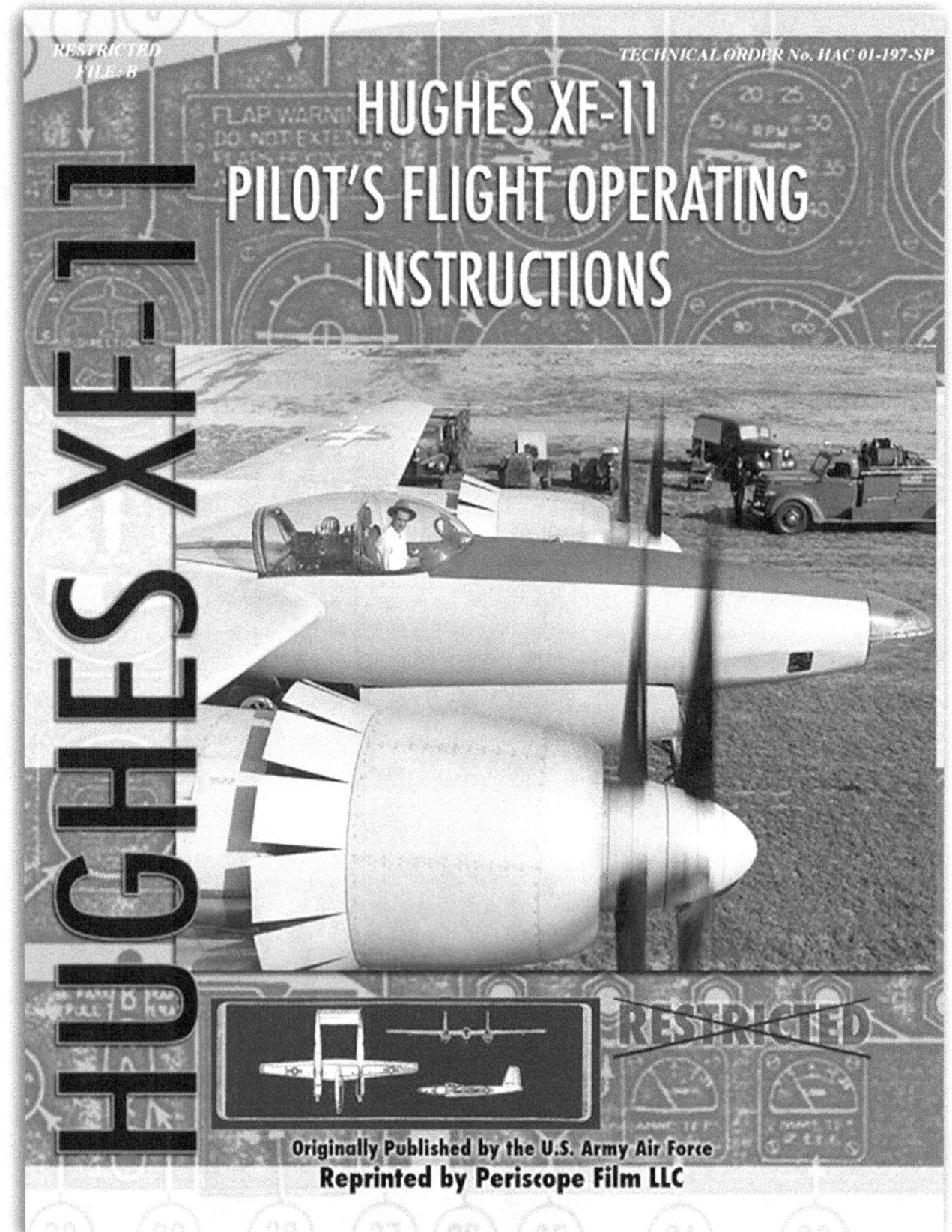

NOW AVAILABLE!

Aircraft At War DVD Series

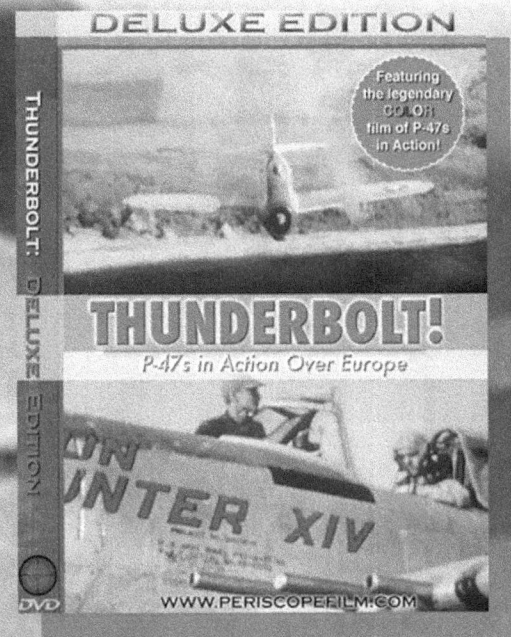
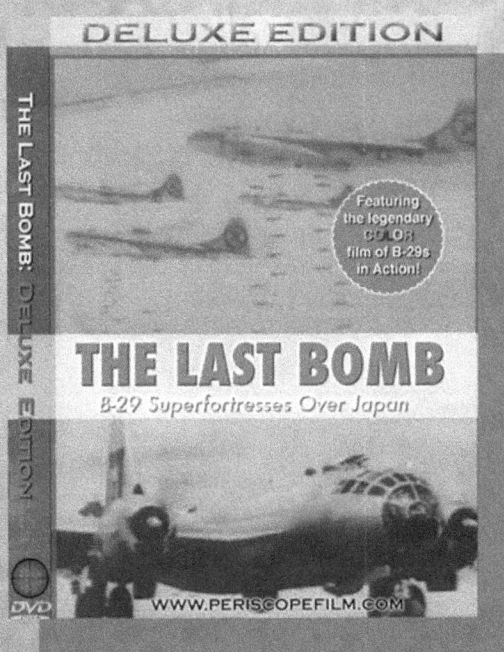
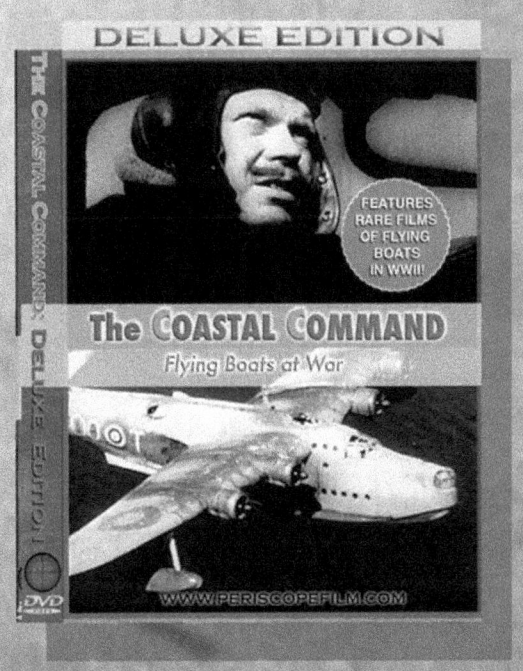

Now Available!

Warships DVD Series

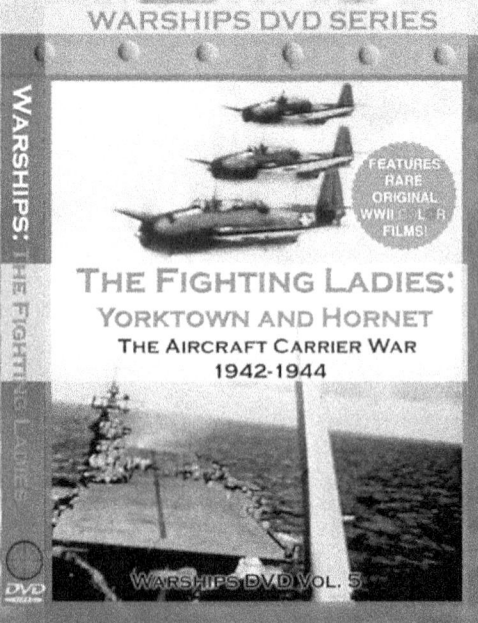

Now Available!

SPRUCE GOOSE
HUGHES FLYING BOAT MANUAL

RESTRICTED

Originally Published by the War Department
Reprinted by Periscope Film LLC

NOW AVAILABLE!

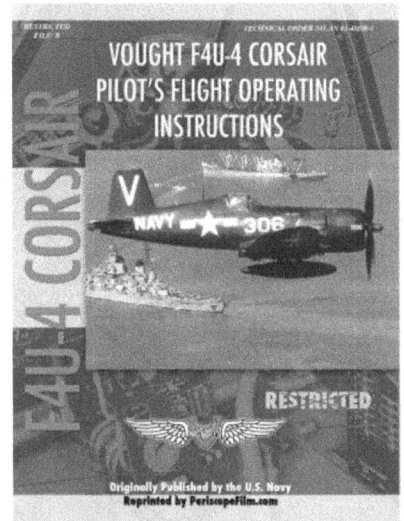
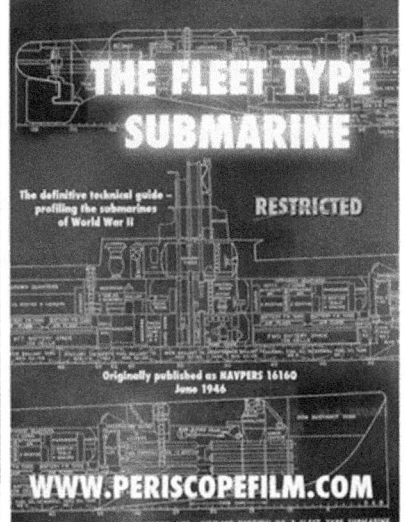

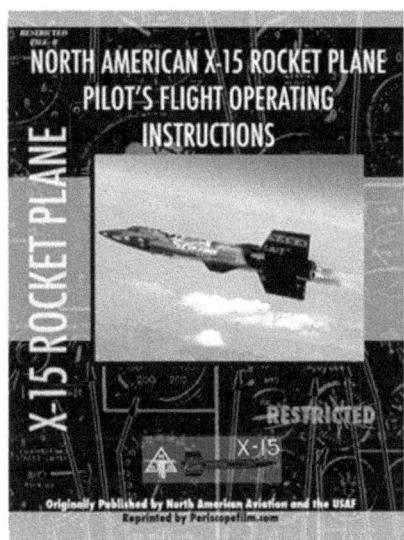

ALSO NOW AVAILABLE FROM PERISCOPEFILM.COM

©2006-2010 Periscope Film LLC
All Rights Reserved
ISBN #978-1-935327-86-8   1-935327-86-0

www.ingramcontent.com/pod-product-compliance
Lightning Source LLC
Chambersburg PA
CBHW081917180426
43199CB00036B/2755